NEUROSCIENCE INTELLIGENCE UNIT

ANTISENSE STRATEGIES FOR THE STUDY OF RECEPTOR MECHANISMS

Robert B. Raffa, Ph.D.
The R.W. Johnson Pharmaceutical Research Institute
Spring House, Pennsylvania, U.S.A.

Frank Porreca, Ph.D.
University of Arizona Health Sciences Center
Tucson, Arizona, U.S.A.

Springer
New York Berlin Heidelberg London Paris
Tokyo Hong Kong Barcelona Budapest

R.G. Landes Company
Austin

NEUROSCIENCE INTELLIGENCE UNIT
ANTISENSE STRATEGIES FOR THE STUDY OF RECEPTOR MECHANISMS

R.G. LANDES COMPANY
Austin, Texas, U.S.A.

International Copyright © 1996 Springer-Verlag, Heidelberg, Germany

All rights reserved.
No part of this book may be reproduced or transmitted in any form or by any means, electronic or mechanical, including photocopy, recording, or any information storage and retrieval system, without permission in writing from the publisher.
Printed in the U.S.A.

Please address all inquiries to the Publishers:
R.G. Landes Company, 909 Pine Street, Georgetown, Texas, U.S.A. 78626
Phone: 512/ 863 7762; FAX: 512/ 863 0081

International distributor (except North America):
Springer-Verlag GmbH & Co. KG
Tiergartenstrasse 17, D-69121 Heidelberg, Germany

 Springer

International ISBN: 3-540-61154-1

While the authors, editors and publisher believe that drug selection and dosage and the specifications and usage of equipment and devices, as set forth in this book, are in accord with current recommendations and practice at the time of publication, they make no warranty, expressed or implied, with respect to material described in this book. In view of the ongoing research, equipment development, changes in governmental regulations and the rapid accumulation of information relating to the biomedical sciences, the reader is urged to carefully review and evaluate the information provided herein.

Library of Congress Cataloging-in-Publication Data

Antisense strategies for the study of receptor mechanisms / Robert B. Raffa, Frank Porreca [editors]
  p. cm. -- (Neuroscience intelligence unit)
    Proceedings of a symposium held in Scottsdale, Arizona from June
      10-15, 1995.
    Includes bibliographical reference and index.
    ISBN 1-57059-345-0 (alk. paper), ISBN 0-412-11331-7
  1. Drug receptors-- Research--Methodology--Congresses.
    2. Antisense nucleic acids--Congresses.   3. Neurotransmitter receptors--Research--Methodology--Congresses.   I. Raffa, Robert B.   II. Porreca, Frank.   III. Series.
RM301.41.A58 1996   96-17121
615'.7--dc20   CIP

Publisher's Note

R.G. Landes Company publishes six book series: *Medical Intelligence Unit, Molecular Biology Intelligence Unit, Neuroscience Intelligence Unit, Tissue Engineering Intelligence Unit, Biotechnology Intelligence Unit* and *Environmental Intelligence Unit.* The authors of our books are acknowledged leaders in their fields and the topics are unique. Almost without exception, no other similar books exist on these topics.

Our goal is to publish books in important and rapidly changing areas of bioscience and environment for sophisticated researchers and clinicians. To achieve this goal, we have accelerated our publishing program to conform to the fast pace in which information grows in bioscience. Most of our books are published within 90 to 120 days of receipt of the manuscript. We would like to thank our readers for their continuing interest and welcome any comments or suggestions they may have for future books.

Deborah Muir Molsberry
Publications Director
R.G. Landes Company

CONTENTS

1. **Antisense "Knockdown" Strategies in Neurotransmitter Receptor Research** 1
   Claes Wahlestedt
   Introduction 1
   Methodological Considerations 3
   Concluding Remarks 6

2. **Antisense Mapping and Opioid Pharmacology** 11
   Gavril W. Pasternak
   Introduction 11
   Effects of Antisense Oligodeoxynucleotides on DOR-1 in NG108-15 Cells and In Vivo 13
   Effect of Antisense Oligodeoxynucleotides Against DOR-1 and KOR-1 In Vivo 15
   Antisense Mapping MOR-1 16
   Antisense Mapping KOR-3 17
   Implications for the Future 19

3. **Antisense Targeting of Delta Opioid Receptors** 25
   Josephine Lai and Frank Porreca
   Antisense Oligodeoxynucleotide Mediated Gene Targeting ... 25
   Pharmacological Evidence for Subtypes of the δ Opioid Receptor ... 26
   Sequence-Specific Blockade of Delta Opioid Receptor Antinociception In Vivo 28
   Sequence-Specific "Knock-Down" of δ Opioid Receptors in Neuroblastoma Cells 32
   Conclusions 33

4. **Functional Effects of Antisense Oligodeoxynucleotides to Opioid Receptors in Rats** 37
   Jill U. Adams, Xiao-Hong Chen, J. Kim DeRiel, Jinling Yin, Martin W. Adler and Lee-Yuan Liu-Chen
   Antisense Oligo Treatment 38
   Antinociception Studies 39
   Thermoregulation Studies 44
   Radioligand Binding 46
   Conclusions 47

5. **In Vivo Antisense Strategy for the Study of Second-Messengers: Application to G-Proteins** 53
   Robert B. Raffa
   Introduction 53
   G-Protein Antisense 55

Experimental Designs 56
   G-Proteins and Antinociception 58
   In Vivo G-Protein Antisense 58
   Caveats 63
   Conclusions and Future Directions 67

6. **Dopamine Antisense Oligodeoxynucleotides
   as Potential Novel Tools for Studying Drug Abuse** 71
   Benjamin Weiss, Long-Wu Zhou and Sui-Po Zhang
   Introduction 71
   Antisense Oligodeoxynucleotides—A Novel Tool
      for Inhibiting the Expression of Dopamine Receptors 73
   D1 Antisense Inhibits D1 Receptor-Mediated Behaviors 75
   Recovery of SKF 38393 Induced-Grooming Behavior
      After Cessation of D1 Antisense Treatment 75
   D1 Antisense Inhibits Rotational Behavior
      Induced by SKF 38393 78
   D1 Antisense Decreases D1 Receptors In Mouse Brain 80
   D3 Antisense Increases Locomotor Behavior
      Induced by Quinpirole 81
   Possible Mechanisms by Which Dopamine Receptor
      Antisense Oligodeoxynucleotides May Be Used
      to Study or Treat Drug Abuse 81
   Summary 85

7. **Electrophysiological Correlates of In Vivo Antisense
   Knockout of Dopamine D2 Autoreceptors
   on Substantia Nigra Dopaminergic Neurons** 93
   James M. Tepper, Bao-C. Sun, Lynn P. Martin and Ian Creese
   Introduction 93
   Autoreceptor Modulation of Dopaminergic Neuronal Activity 94
   Methods 96
   Results 98
   Discussion 102

8. **C-Fos Antisense: Infusion and Effects** 111
   Deborah Young and Michael Dragunow
   Introduction 111
   Immediate Early Genes 112
   Antisense Oligonucleotides 112
   The In Vivo Use of *c-Fos* Antisense 114
   Future Directions 122

9. Antisense Strategy and Stimulus-Secretion Coupling 135
Marjan Rupnik and Robert Zorec
Introduction 135
Mechanism(s) of Action of Antisense Oligonucleotides
 on Protein Expression 136
Delivery of Antisense Oligonucleotides into Cytosol 139
Application of Antisense Strategy to Study the Modulation
 and Function of Voltage-Activated Calcium Channels 144
Antisense Strategy and the Assignment of Monomeric
 and Heterotrimeric GTP-Binding Proteins
 in the Secretory Activity 145
Summary and Perspectives 147

10. Antisense Targeting of Corticotropin-Releasing Hormone and Corticotropin-Releasing Hormone Receptor Type I 153
Thomas Skutella, Joseph Christopher Probst, Christian Behl and Florian Holsboer
CRH and Anxiety 153
Selection of CRH Antisense ODNs 155
CRH Antisense ODNs and Anxiety 158
CRH Receptor (Type I) and Anxiety 163
Conclusion 167

11. Application of Antisense Strategies for the Analysis of Protein Kinase Functions 175
Jesús Avila, Javier Díaz-Nido and Nieves Villanueva
Introduction 175
Protein Kinase 2 CK2 (Casein Kinase 2) 176
cAMP-Dependent Protein Kinase 178
PKC 180
Proline-Directed Protein Kinases (PDPK) 181
Tyrosine Protein Kinases 183
Phosphatases 183
Concluding Remarks 184

12. Application of Antisense Technology for Studying the Functional Role of Tau Proteins in Neural Plasticity 189
Maurizio Memo
Introduction 189
Tau Proteins in Neurons Developing In Vitro 191
Tau Antisense in Neurons Developing In Vitro 192
Tau in Mature and Degenerating Neurons 192
Tau Antisense in Degenerating Neurons 194
Conclusion 195

13. **Behavioral Assessment of Antisense Oligonucleotides Targeted to Messenger RNAs of Genes Associated with Alzheimer's Disease** 201
David P. Binsack, Sudhir Agrawal and Charles A. Marotta
   Introduction 201
   Alzheimer's Disease 202
   Amyloid Precursor Protein and β-Amyloid 202
   Apolipoprotein E 203
   Antisense Oligonucleotides 204
   Spatial Tasks Designed to Test Learning and Memory 205
   Water Maze Paradigm for Assessing Learning and Memory 205
   Sensitivity of Water Maze to Compounds
      with Anticholinergic Activity 208
   Conditions for Testing Anti-APP and Anti-ApoE ASOs 211
   Future Studies 216

Index 221

EDITORS

Robert B. Raffa, Ph.D.
Research Fellow
The R.W. Johnson Pharmaceutical Research Institute
Spring House, Pennsylvania, U.S.A.
Chapter 5

Frank Porreca, Ph.D.
Professor, Department of Pharmacology
University of Arizona Health Sciences Center
Tucson, Arizona, U.S.A.
Chapter 3

CONTRIBUTORS

Jill U. Adams, Ph.D.
Temple University School
   of Medicine
Philadelphia, Pennsylvania, U.S.A.
Chapter 4

Martin W. Adler, Ph.D.
Temple University School
   of Medicine
Philadelphia, Pennsylvania, U.S.A.
Chapter 4

Sudhir Agrawal, Ph.D.
Hybridon, Inc.
Worcester, Massachusetts, U.S.A.
Chapter 13

Jesús Avila, Ph.D.
Universidad Autonoma de Madrid
Madrid, Spain
Chapter 11

Christian Behl, Ph.D.
Max Planck Institute of Psychiatry
Munich, Germany
Chapter 10

David P. Binsack, Ph.D.
Brown University
Providence, Rhode Island, U.S.A.
Chapter 13

Xiao-Hong Chen, M.D.
Temple University School
   of Medicine
Philadelphia, Pennsylvania, U.S.A.
Chapter 4

Ian Creese, Ph.D.
Rutgers, The State University
   of New Jersey
Newark, New Jersey, U.S.A.
Chapter 7

J. Kim DeRiel, Ph.D.
Philadelphia, Pennsylvania, U.S.A.
Chapter 4

Javier Díaz-Nido, Ph.D.
Universidad Autonoma de Madrid
Madrid, Spain
Chapter 11

Michael Dragunow, Ph.D.
University of Auckland School
  of Medicine
Auckland, New Zealand
Chapter 8

Florian Holsboer, M.D., Ph.D.
Max Planck Institute of Psychiatry
Munich, Germany
Chapter 10

Josephine Lai, Ph.D.
University of Arizona Health Sciences
  Center
Tucson, Arizona, U.S.A.
Chapter 3

Lee-Yuan Liu-Chen, Ph.D.
Temple University School
  of Medicine
Philadelphia, Pennsylvania, U.S.A.
Chapter 4

Charles A. Marotta, M.D., Ph.D.
Brown University
Providence, Rhode Island, U.S.A.
Chapter 13

Lynn P. Martin, Ph.D.
Rutgers, The State University
  of New Jersey
Newark, New Jersey, U.S.A.
Chapter 7

Maurizio Memo, Ph.D.
University of Brescia School
  of Medicine
Brescia, Italy
Chapter 12

Gavril W. Pasternak, M.D., Ph.D.
Memorial Sloan-Kettering Cancer
  Center
New York, New York, U.S.A.
Chapter 2

Joseph Christopher Probst, Ph.D.
Max Planck Institute of Psychiatry
Munich, Germany
Chapter 10

Marjan Rupnik, Ph.D.
University of Ljubljana School
  of Medicine
Ljubljana, Slovenia
Chapter 9

Thomas Skutella, M.D.
Institute of Anatomy
Humboldt University
Berlin, Germany
Chapter 10

Bao-Cun Sun, Ph.D.
Rutgers, The State University
  of New Jersey
Newark, New Jersey, U.S.A.
Chapter 7

James M. Tepper, Ph.D.
Rutgers, The State University
  of New Jersey
Newark, New Jersey, U.S.A.
Chapter 7

Nieves Villanueva, Ph.D.
Instituto de Salyd Carlos III
Majadahonda, Spain
Chapter 11

Claes Wahlestedt, M.D., Ph.D.
Astra Pain Research Unit
Laval (Montreal), Quebec, Canada
Chapter 1

Benjamin Weiss, Ph.D.
Medical College of Pennsylvania
 and Hahnemann University
Philadelphia, Pennsylvania, U.S.A.
Chapter 6

Jinling Yin, B.S.
Temple University School
 of Medicine
Philadelphia, Pennsylvania, U.S.A.
Chapter 4

Deborah Young, Ph.D.
University of Auckland School
 of Medicine
Auckland, New Zealand
Chapter 8

Sui-Po Zhang
Medical College of Pennsylvania
 and Hahnemann University
Philadelphia, Pennsylvania, U.S.A.
Chapter 6

Long-Wu Zhou
Medical College of Pennsylvania
 and Hahnemann University
Philadelphia, Pennsylvania, U.S.A.
Chapter 6

Robert Zorec, Ph.D.
University of Ljubljana School
 of Medicine
Ljubljana, Slovenia
Chapter 9

PREFACE

The idea for this book came about following a Symposium that we co-chaired at the annual meeting of the College on Problems of Drug Dependence (CPDD) held at the Scottsdale Princess Hotel in Scottsdale, Arizona from June 10-15, 1995. The title of the Symposium was "Application of Antisense Strategies for Investigation of Receptor Mechanisms In vivo and In vitro." This was the first time that a Symposium specifically devoted to the topic of an in vivo antisense strategy was held for the CPDD. Partly because of the novelty of the in vivo antisense approach itself, and partly because of the vast variety of areas of research for which such an approach can provide valuable information and insight, there was a large interest in the Symposium—as judged by both the number of attendees and by the number and extent of the questions. From this interest, arose the collective feeling that the availability in written form of as much of the material presented at the Symposium as possible would be of value. It was further felt that the work of other investigators who also use the in vivo antisense strategy, but were not present at the Symposium, should also be collected into a single source. We are happy to say that invitations to these authors were gratefully accepted and their work constitutes an important part of this book. The material presented by the Symposium speakers and by the invited authors provides, we believe, the "state-of-the-art" of the in vivo antisense strategy and an indication of the breadth of its applicability. As such, it should serve as a valuable resource to the general reader (who is interested in learning about the strategy and the potential implications and opportunities for their own work) as well as for those who wish to use the techniques outlined by each author for specific applications. Each of the chapters has been designed in such a way that it can be read independently of the others, but has been written with a uniformity of theme and style that should allow smooth transitions. In addition, a general introduction to the concepts of antisense, and its in vivo use, has been provided for those new to this field. It is our hope that this book provides the reader with an opportunity to quickly learn about the technique and to review the types of applications for which it has already been used. We also hope that it serves as an impetus to new work in this area. As with any new field of research, there are inherent difficulties of application or interpretation. The in vivo antisense strategy is still in its infancy and with maturity will come grace. The material contained in this book is supplied in order to give the reader an indication of the power of the technique, the versatility of its application and the insight that it might provide for their own work.

Robert B. Raffa, Ph.D.
Frank Porreca, Ph.D.

CHAPTER 1

ANTISENSE "KNOCKDOWN" STRATEGIES IN NEUROTRANSMITTER RECEPTOR RESEARCH

Claes Wahlestedt

INTRODUCTION

Although it is not a new concept, the research concerning antisense oligodeoxynucleotide (ODN) inhibition of gene expression has gained considerable momentum in the past few years. These techniques are today widely used by preclinical researchers and in addition, there are significant efforts to develop antisense compounds as therapies for cancer, AIDS, and other diseases. In the late 1970s, Zamecnik and coworkers[1] described specific inhibition of gene expression in cell cultures using antisense ODNs. Since then, several modifications of the techniques have been tested in a variety of biological systems. Great emphasis has been placed on chemical modifications of the phosphodiester ODNs, resulting in greater stability. Today, the field is heavily dependent on phosphorothioate analogs which unfortunately are associated with a number of nonspecific actions. ODNs have been shown to be specific inhibitors of protein expression. Presumably, through a variety of mechanisms including binding via Watson-Crick base pairs to target RNA in an antisense orientation or binding to a targeted gene or duplex DNA by triple helix formation. In addition, there is much evidence, in particular from the extensive use of phosphorothioates, supporting antisense ODN inhibition of protein function by direct binding in what has been referred to as aptamer binding.

Antisense Strategies for the Study of Receptor Mechanisms,
edited by Robert B. Raffa and Frank Porreca. © 1996 R.G. Landes Company.

Our research group started antisense work around 1990 at a time when to the best of our knowledge, nobody had targeted receptors, let alone in the nervous system. Today, the list of targeted receptors is lengthy (see Table 1.1). Our reason for taking this approach was to enable us to study receptors for which conventional antagonists were

Table 1.1. Some examples from the literature of antisense oligodeoxynucleotide inhibition of receptor expression. The targeted receptors are listed alphabetically

Receptor	Reference
Angiotensin 1	**Sakai et al 1994,**[2]* **Meng et al 1994**[3]
Bradykinin B2	Webb et al 1994[4]
Cholecystokinin B	**Vanderah et al 1994**[5]
Delta-opioid	**Standifer et al 1994,**[6] **Lai et al 1994**[7] **Bilsky et al 1994,**[8] **Tseng et al 1994**[9]
Dopamine-D1	Zhang et al 1994[10]
Dopamine-D2	**Zhang and Creese 1993,**[11] **Zhou et al 1994**[12] **Silvia et al 1994,**[13] **Valerio et al 1994**[14]
Endothelin (ET-A)	Adner et al 1994[15]
Estrogen	**McCarthy et al 1993,**[16] **Xu and Thomas 1994**[17]
GABA-B	Holopainen and Wojcik 1993[18]
Gastrin-Releasing Peptide	Bitar and Zhu, 1993[19]
Glutamate NMDA-R1	**Wahlestedt et al 1993,**[20] **Soltesz et al 1994**[21]
NMDA-R2A	Bessho et al 1994[22]
delta-2	Hirano et al 1994[23]
Glu-R1	Vanderklish et al 1992[24]
Insulin-like growth factor I	Reiss et al 1992,[25] Resnicoff et al 1993,[26] Wada et al 1993[27]
Interleukin-1	Burch and Mahan 1991[28]
Interleukin-6	Levy et al 1991[29]
Interleukin-8	Ishiko et al 1995[30]
Kappa-opioid	**Chien et al 1994,**[31] **Adams et al 1994**[32]
Luteinizing hormone	Cooke and West 1992[33]
Mu-opioid	Rossi et al 1994[34]
Muscarinic acetylcholine	Holopainen and Wojcik 1993,[18] **Zhang et al 1994**[35]
Neuromedin B	Bitar and Zhu 1993[19]
Neuropeptide Y-Y1	**Wahlestedt et al 1992,**[36] **1993,**[36a] **Erlinge et al 1993**[37]
Nicotinic acetylcholine	Listerud et al 1991,[38] Brussaard et al 1994[39]
Oxytocin	McCarthy et al 1994[16]
p75 nerve growth factor	Barrett and Bartlett 1994[40]
Progesterone	**Pollio et al 1993,**[41] **Mani et al 1994,**[42] **Ogawa et al 1994**[43]
Substance P (NK-1)	Ogo et al 1994[44]
Thyrotropin-releasing hormone	Matusleibowitz et al 1994[45]
Transferrin	Sasaki et al 1993,[46] Kato et al 1994[47]
Urokinase	Quattrone et al 1993,[48] Gyetko et al 1994[49]

*References in bold indicate in vivo treatment

not available, or where such agents showed limited selectivity. After having found that cultured cells (primary cultures of neurons or vascular smooth muscle cells) readily take up labeled ODNs from media, we addressed the question whether receptor expression could be inhibited by ODNs in vitro and, subsequently, in vivo. The present review is primarily aimed at giving advice to the novice interested in starting up antisense ODN work. Most emphasis will be given to the potential use of the technique in studies of the living brain.

METHODOLOGICAL CONSIDERATIONS

IN VITRO TESTING

Whenever possible it is advisable to test ODN efficacy in vitro prior to in vivo experimentation (Table 1.1). Concentration response curves of ODNs typically are steep and it is therefore important to study a narrow range of ODN concentrations. Unmodified or phosphorothioated ODNs have often been found to be effective at ~1 µM. Mismatched ODNs (see below) should be tested at the same concentrations as the antisense ODNs, allowing the elucidation of a window in which the active ODN is void of sequence independent actions.

An important issue regarding in vitro experiments is whether serum which contains nuclease activities can be omitted or not. This is possible with many types of primary cultures and some cell lines as well as tissue samples such as blood vessel segments.[50] In our experience, even heat inactivated sera will degrade unmodified ODNs fairly rapidly. In serum free conditions the ODNs are, however, relatively stable at 37°C and there is only a modest need for adding additional ODNs during the course of the experiment.

Over the past few years, it has become quite clear that cultured cells vary widely in their ability to take up ODN from the (extracellular)

Table 1.2. Some experimental considerations for studies of antisense oligodeoxynucleotide (ODN) inhibition of protein expression in living brain (See the text for further details)

1. In vitro → in vivo testing
2. Stringent mismatched ODN controls
3. Two or more active antisense ODNs
4. Careful screening of ODN binding sites
5. Phosphodiester ODN → partially substituted → fully substituted phosphorothioate ODN
6. Repeated or continuous ODN application
7. CSF or site injection
8. Steep ODN concentration-effect relationships
9. Behavior/biochemistry

media. In our experience,[51] some primary cultured cells, such as neurons, will readily take up ODN. In sharp contrast, all cell lines we routinely use for transfection experiments (e.g., CHO, HEK-293, COS, HeLa) show very low uptake of ODN. This may perhaps be due to lower pinocytotic activities in the latter cells which have been passaged numerous times. Studies to directly support the latter suggestion have yet to be carried out.

Mismatched ODN Analog Control

Typically investigators will use antisense ODNs 15-22 nucleotides in length. As a minimum requirement, they must include a mismatched analog, where 1-4 nucleotides are scrambled. On average each mismatch will correspond to a 500-fold reduction of hybridization affinity.[51] However, there is a wide variability with certain mismatches being virtually ineffective whereas other single mismatches will abolish activity in a given test systems. The mismatched ODN (the fewer mismatches the better) represents a more stringent control as compared to scrambled or sense ODN controls. Thus, investigators should always include a mismatched control ODN in their antisense experiments. The latter holds true also for "pilot" experiments where perhaps only two experimental groups, one with antisense ODN and one with mismatch ODN, are used.

Multiple Antisense Control

When targeting a given protein for the first time, it is reassuring to find inhibition by two or more antisense sequences interacting with different sites of the mRNA (Table 1.1).

Screening ODN Binding Sites

A number of approaches can be taken. In general, antisense sequences should not have undesired secondary structures nor should the target mRNA region contain such sequences. Typically, however, a number of antisense ODNs are synthesized and efficacy is tested. In our own work, less than 20% of tested antisense ODNs have shown sufficient antisense efficacy to warrant further characterization.

Choice of Analog

Caution must be taken when using the commonly available phosphorothioate analog which is known to produce sequence-nonspecific actions (Table 1.3). This is not the case for unmodified (phosphodiester) ODNs. Endprotected phosphodiester ODNs as well as other partially phosphorothioated analogs appear to show reduced toxicity. However, there is still a lack of systematic studies investigating mixed analogs in any assay, let alone the living brain.

Among additional analogs, two types deserve mention. The peptide nucleic acids (PNAs) are theoretically very interesting but have

Table 1.3. Phosphorothioate oligodeoxynucleotides produce sequence independent actions

Action	Reference
protein binding	Brown et al 1994[52]
non-target specific RNAse cleavage	Weidner and Busch 1994[53]
induction of Sp1 transcription	Perez et al 1994[54]
binding to nucleolar protein C23.nucleolin	Weidner et al 1995[55]
brain lesions	Hooper et al 1994[56]
chronic stress-like activation on HPA axis	Skutella et al 1994[57]
but: partially phosphorothioated analogs may show better profiles	Zhao et al 1993,[58] Erlich et al 1994[59] Robertson 1995[60]

not undergone sufficient biological testing.[61] A drawback of PNAs may, however, be their poor cellular uptake. A much older analog than PNA is the methylphorphonate. The latter molecule lacks the appropriate charges to become a substrate for RNase H when bound to mRNA and may, although sufficiently nuclease resistant, for this reason be less efficacious than phosphorothioate and phosphodiester ODNs (see also ref. 1).

Mode of Brain Administration

ODNs can be administered either into the cerebrospinal fluid (CSF), e.g., into lateral ventricle or intrathecal space, or into specific sites of the brain parenchyma. Care should obviously be taken not to induce brain damage. As stated in the above, phosphodiesters can successfully be used in the brain as they are essentially stable in CSF. The latter type of molecule will upon CSF administration reach the parenchyma by way of diffusion as well as by passing through perivascular spaces and subarachnoid space.[62] The ODNs can be administered either repeatedly or by continuous infusion, e.g., by osmotic minipumps. Repeated injections are typically carried out twice daily for some 2-7 days when studying constitutively expressed proteins such as receptors. As stated above, in vitro concentrations of 1 µM are typically required to obtain inhibition of protein expression. Our original in vivo experiments were designed such that similar or higher CSF concentrations could be obtained over several hours considering elimination by CSF bulk flow.[51]

Biochemistry and Function

When observing a functional consequence of antisense ODN application to the living brain, e.g., altered behavior, the tissue should be saved for appropriate biochemical analysis. In receptor studies, reductions in Bmax are then to be anticipated. It is important to analyze also another (control) receptor, preferably a subtype of the one being targeted. In addition, general toxicity in brain tissue should be studied biochemically as well as histologically.[51]

CONCLUDING REMARKS

In the past few years, many receptors have been targeted with antisense ODNs (Table 1.1). Their uses have been either as tools in drug target validation or as potential new drugs, e.g., as an alternative to conventional receptor antagonists in cases when such agents have been lacking or shown limited selectivity. It is anticipated that the use of antisense ODNs for protein "knockdown" will continue to rise, particularly until conditional "knock-out" animals can be produced reliably.

REFERENCES

1. Zamecnik PC. Oligonucleotide base hybridization as a modulator of genetic message readout. In: Wickstrom E, ed. Prospects for Antisense Nucleic Acid Therapy of Cancer and AIDS. New York: Wiley-Liss, 1991:1.
2. Sakai RR, He PF, Yang XD et al. Intracerebroventricular administration of AT1 receptor antisense oligonucleotides inhibits the behavioral actions of angiotensin II. J Neurochem 1994; 62: 2053-2056.
3. Meng HB, Wielbo D, Gyurko R et al. Antisense oligodeoxynucleotide to AT(1) receptor mRNA inhibits central angiotensin induced thirst and vasopressin. Reg Peptides 1994; 54:543-551.
4. Webb M, McIntyre P, Phillips E. B1 and B2 bradykinin receptors encoded by distinct mRNAs. J Neurochem 1994; 62:1247-1253.
5. Vanderah TW, Lai J, Yamamura HI et al. Antisense oligodeoxynucleotide to the CCKB receptor produces naltrindole- and [leu(5)]enkephalin antiserum-sensitive enhancement of morphine antinociception. Neuroreport 1994; 5:2601-2605.
6. Standifer KM, Chien CC, Wahlestedt C et al. Selective loss of delta opioid analgesia and binding by antisense oligodeoxynucleotides to a delta opioid receptor. Neuron 1994; 12:805-810.
7. Lai J, Bilsky EJ, Rothman RB et al. Treatment with antisense oligonucleotide to the opioid delta receptor selectively inhibits delta-2 agonist antinociception. Neuroreport (England) 1994; 5:1049-1052.
8. Bilsky EJ, Bernstein RN, Pasternack GW et al. Selective inhibition of [D-Ala2, Glu4]deltorphin antinociception by supraspinal, but not spinal, administration of an antisense oligodeoxynucleotide to an opioid delta receptor. Life Sci (England) 1994; 55:PL37-43.
9. Tseng LF, Collins KA et al. Antisense oligodeoxynucleotide to a delta opioid receptor selectively blocks the spinal antinociception induced by delta- but not mu- or kappa-opioid receptor agonists in the mouse. Eur J Pharmacol 1994; 258:R1-3.

10. Zhang SP, Zhou LW, Weiss B. Oligodeoxynucleotide antisense to the D-1 dopamine receptor mRNA inhibits D-1 dopamine receptor-mediated behaviors in normal mice and in mice lesioned with 6-hydroxydopamine. J Pharmacol Exp Ther 1994; 271:1462-1470.
11. Zhang M, Creese I. Antisense oligodeoxynucleotide reduces brain dopamine D2 receptors: behavioral correlates. Neurosci Lett (Ireland) 1993; 16:332-226.
12. Zhou LW, Zhang, SP, Qin ZH et al. In vivo administration of an oligodeoxynucleotide antisense to the D2 dopamine receptor messenger RNA inhibits D2 dopamine receptor-mediated behavior and the expression of D2 dopamine receptors in mouse striatum. J Pharmacol 1994; 268:1015-1023.
13. Silvia CP, King GR, Lee TH et al. Intranigral administration of D2 dopamine receptor antisense oligodeoxynucleotides establishes a role for nigrostriatal D2 autoreceptors in the motor neuron actions of cocaine. Mol Pharmacol 1994; 46:51-57.
14. Valerio A, Alberici A, Tinti C et al. Antisense strategy unravels a dopamine receptor distinct from the D2 subtype, uncoupled with adenyl cyclase, inhibiting prolactin release from rat pituitary cells. J Neurochem 1994; 62:1260-1266.
15. Adner M, Erlinge D, Salford LG et al. Human endothelin ET(A) receptor antisense oligodeoxynucleotides inhibit endothelin-1 evoked vasoconstriction. Eur J Pharmacol 1994; 251:281-284.
16. McCarthy MM, Kleopoulos SP, Mobbs CV et al. Infusion of antisense oligodeoxynucleotides to the oxytocin receptor in the ventromedial hypothalamus reduces estrogen-induced sexual receptivity and oxytocin receptor binding in the female rat. Neuroendocrinology 1994; 59:432-440.
17. Xu XM, Thomas ML. Estrogen receptor-mediated direct stimulation of colon cancer cell growth in vitro. Mol Cell Endocrin 1994; 105:197-201.
18. Holopainen I, Wojcik WJ. A specific antisense oligonucleotide to mRNAs encoding receptors with seven transmembrane spanning regions decreases muscarinic m2 and gamma-aminobutyric acidB receptors in rat cerebellar granule cells. J Pharmacol Exp Ther 1993; 264:423-430.
19. Bitar KN, Zhu XX. Expression of bombesin-receptor subtypes and their differential regulation of colonic smooth muscle contraction. Gastroenterology 1993; 105:1672-1680.
20. Wahlestedt C, Golanov E, Yamamoto S et al. Antisense oligodeoxynucleotides to the NMDAR1 receptor channel protect cortical neurones from excitotoxicity and reduce focal ischemic infarctions in rat. Nature 1993; 363:260-262.
21. Soltesz I, Zhou Z, Smith GM et al. Rapid turnover rate of the hippocampal synaptic NMDA-R1 receptor subunits. Neuroscience Letters 1994; 181:5-8.
22. Bessho Y, Nawa H, Nakanishi S. Selective uptake of an NMDA receptor subunit mRNA in cultured cerebellar granule cells by K(+)-induced depolarization and NMDA treatment. Neuron 1994; 12:87-95.

23. Hirano T, Kasono K, Araki K et al. Involvement of the glutamate receptor delta-2 subunit in the long-term depression of glutamate responsiveness in cultured rat purkinje cells. Neuroscience Letters 1994; 182:172-176.
24. Vanderklish P, Neve R, Bahr BA et al. Translational suppression of a glutamate receptor subunit impairs long-term potentiation. Synapse 1992; 13:333-337.
25. Reiss K, Porcu P, Sell C et al. The insulin-like growth factor 1 receptor is required for the proliferation of hemopoeitic cells. Oncogene (England) 1992; 7:2243-2248.
26. Resnicoff M, Ambrose D, Coppola D et al. Insulin-like growth factor-1 and its receptor mediate the autocrine proliferation of human ovarian carcinoma cell lines. Lab Invest (United States) 1993; 69:756-760.
27. Wada J, Liu ZZ, Alvares K. Cloning of cDNA for the alpha subunit of mouse insulin-like growth factor I receptor and the role of the receptor in metanephric development. Proc Natl Acad Sci 1993; 90:10360-10364.
28. Burch RM, Mahan LC. Oligonucleotides antisense to the interleukin 1 receptor mRNA block the effects of interleukin 1 in cultured murine and human fibroblasts and in mice. J Clin Invest 1991; 88(4):1190-1196.
29. Levy Y, Tsapis A, Brouet JC. Interleukin-6 antisense oligonucleotides inhibit the growth of human myeloma cell lines. J Clin Invest 1991; 88:696-699.
30. Ishiko T, Sakamoto K, Yamashita SI et al. Carcinoma cells express IL-8 and the IL-8 receptor—their inhibition attenuates the growth of carcinoma cells. Int J Oncol 1995; 6:119-122.
31. Chien CC, Brown G, Pan YX et al. Blockage of U50, 488H analgesia by antisense oligodeoxynucleotides to a kappa-opioid receptor. Eur J Pharmacol 1994; 253:R7-8.
32. Adams JU, Chen XH, Deriel JK et al. Intracerebroventricular treatment with an antisense oligodeoxynucleotide to kappa-opioid receptors inhibited kappa-agonist-induced analgesia in rats. Brain Research 1994; 667:129-132.
33. Cooke BA, West AP. Investigations into the structure-activity relationships of the luteinizing hormone receptor using antisense oligodeoxynucleotides. Biochem Soc Trans 1992; 20:754-756.
34. Rossi C, Pan YX, Cheng J et al. Blockade of morphine analgesia by an antisense oligodeoxynucleotide.
35. Zhang SP, Zhou LW, Weiss B. Oligodeoxynucleotide antisense to the D-1 dopamine receptor mRNA inhibits D-1 dopamine receptor-mediated behaviors in normal mice and in mice lesioned with 6-hydroxydopaminer. J Pharmacol Exp Ther 1994; 271:1462-1470.
36. Wahlestedt C, Yee F, Yoo GF et al. Functional consequences of in vitro and in vivo downregulation of brain neuropeptide Y (NPY) Y1-type receptors by antisense oligodeoxynucleotides. In: Zalcman S, Scheller R, Tsien R, eds. Molecular Neurobiology Proceedings of the Second NIMH Conference. Rockville, Maryland: National Institute of Mental Health, 1992:280.

36a. Wahlestedt, C., Pich, E.M., Koob G.F., Yee, F. and Heilig, M.: Modulation of anxiety and neuropeptide Y-Y1 receptors by antisense oligodeoxynucleotides. Science, 1993: 259:528-531
37. Erlinge D, Edvinsson L, Brunkwall J et al. Human neuropeptide Y Y1 receptor antisense oligodeoxynucleotide specifically inhibits neuropeptide Y-evoked vasoconstriction. Eur J Pharmacol 1993; 240:77-80.
38. Listerud M, Brussaard AB, Devay P et al. Functional contribution of neuronal AChR subunits revealed by antisense oligonucleotides. Science 1991; 254:1518-1521.
39. Brussaard AB, Yang X, Doyle JP et al. Developmental regulation of multiple nicotinic ACHR channel subtypes in embryonic chick habenula neurons—contributions of both the alpha-2 and alpha-4 subunit genes. Pflugers Archiv—Eur J Physiol 1994; 429:27-43.
40. Barrett GL, Bartlett PF. The P75 nerve growth factor receptor mediates survival or death depending on the stage of sensory neuron development. Proc Natl Acad Sci 1994; 91:6501-6505.
41. Pollio G, Xue P, Zanisi M et al. Antisense oligonucleotide blocks progesterone-induced lordosis behavior in ovariectomized rats. Brain Res Mol Brain Res 1993; 19:35-39.
42. Mani SK, Blaustein JD, Allen JM et al. Inhibition of rat sexual behavior by antisense oligodeoxynucleotides to the progesterone receptor. Endocrinology 1994; 135:1409-1414.
43. Ogawa S, Olazabal UE, Parhar IS et al. Effects of intrahypothalamic administration of antisense DNA for progesterone receptor mRNA on reproductive behavior and progesterone receptor immunoreactivity in female rat. J Neurosci 1994; 14:1766-1774.
44. Ogo H, Hirai Y, Miki S et al. Modulation of substance P neurokinin-1 receptor in human astrocytoma cells by antisense oligodeoxynucleotides. General Pharmacology 1994; 25: 1131-1135.
45. Matusileibowitz N, Nussenzveig DR, Gershengorn MC et al. The hemispheric functional expression of the thyrotropin-releasing-hormone receptor is not determined by the receptors physical distribution. Biochem J 1994; 303:129-134.
46. Sasaki K, Zak O, Aisen P. Antisense suppression of transferrin receptor gene expression in a human hepatoma cell (HuH-7) line. Am J Hematol 1993; 42:74-80.
47. Kato J, Kohgo Y, Kondo H et al. Antisense oligodeoxynucleotides for IL-2, c-myc and transferrin receptor synchronize mitogen-activated lymphocytes in the G1 phase. Scand J Immunol (England) 1994; 39:499-504.
48. Quattrone A, Fibbi G, Anichini E et al. Reversion of the invasive phenotype of transformed human fibroblasts by anti-messenger oligonucleotide inhibition of urokinase receptor gene expression. Cancer Res 1995; 55:90-95.
49. Gyetko MR, Todd RF, Wilkinson CC et al. The urokinase receptor is required for human monocyte chemotaxis in vitro. Journal of Clinical Investigation 1994; 93:1380-1387.

50. Erlinge D, You J, Wahlestedt C et al. Characterization of an ATP receptor mediating mitogenesis in vascular smooth muscle cells. Eur J Pharmacol 1995; 289:135-149.
51. Wahlestedt C. Antisense oligonucleotide strategies in neuropharmacology. TiPS 1994; 15:42-46.
52. Brown DA, Kang SH, Gryaznov SM et al. Effect of Phosphorothioate modification of oligonucleotides on specific protein binding. Journal of Biological Chemistry 1994; 269:26801-26805.
53. Weidner DA, Busch H. Antisense phosphorothioate oligonucleotides direct both site-specific and nonspecific RNase H cleavage of in vitro synthesized p120 mRNA. Oncology Research 1994; 6:237-242.
54. Perez JR, Stein CA, Majumder S et al. Sequence-independent induction of Sp1 transcription factor activity by phosphorothiate oligonucleotides. Proc Nat Acad Sci 1994; 91: 5957-5961.
55. Weidner DA, Valdez BC, Henning D et al. Phosphorothioate oligonucleotides bind in a non sequence-specific manner to the nucleolar protein C23/nucleolin. FEBS Lett 1995; 366:146-50.
56. Hooper ML, Chiasson BJ, Robertson HA. Infusion into the brain of an antisense oligonucleotide to the immediate-early gene c-fos suppresses production of fos and produces a behavioral effect. Neuroscience 1994; 63:917-24.
57. Skutella T, Stohr T, Probst JC et al. Antisense oligodeoxynucleotides for in vivo targeting of corticotrophin-releasing hormone mRNA—comparison of phosphorothioate and 3'-inverted probe performance. Hormone and Metabolic Research 1994; 26:460-464.
58. Zhao Q, Matson S, Herrera CJ et al. Comparison of cellular binding and uptake of antisense phosphodiester, phosphorothioate, and mixed phosphorothioate and methylphosphonate oligonucleotides. Antisense Res Dev 1993; 3:53-68.
59. Erlich G, Patinkin D, Ginzberg D et al. Use of partially phosphorothioated antisense oligodeoxynucleotides for sequence-dependent modulation of hematopoeisis in culture. Antisense Research & Development 1994; 4:173-183.
60. Robertson HA. Antisense oligodeoxynucleotides: comparison of the uptake and localization of partial and full phosphorothioate derivatives in brain. Soc Neurosci Abstr 1995; 21(3):2080.
61. Nielsen PE. DNA analogues with nonphosphodiester backbones. Ann Rev Biophys Biomol Structure 1995; 24:167-183.
62. Yee F, Ericson H, Reis DJ, Wahlestedt C. Cellular uptake of intracerebroventricularly administered biotin- or digoxigenin-labelled antisense oligodeoxynucleotides in the rat. Cell Mol Neurobiol 1994; 14:475-486.

= CHAPTER 2 =

ANTISENSE MAPPING AND OPIOID PHARMACOLOGY

Gavril W. Pasternak

INTRODUCTION

Opioids are among the oldest drugs in use and their pharmacology has long been the object of study. Detailed and sophisticated pharmacological approaches have proposed a variety of opioid receptors, each with its own pharmacological actions (Table 2.1; see review).[1] Three major classes have been defined and named according to their prototypic ligands. Each one can produce analgesia, but through pharmacologically and regionally distinct mechanisms. Mu receptors, named after morphine, are responsible for many of the side-effects associated with analgesics, including respiratory depression and the inhibition of gastrointestinal transit. Delta receptors are selective for the enkephalins and a number of their derivatives. The kappa family is more complex. Kappa$_1$ receptors have been defined by their high affinity for a series of compounds, such as U50,488H and U69,593. Several U50,488H-insensitive subtypes of kappa receptors also have been reported.[2-4] The kappa$_2$ receptors have been observed in binding assays, but require a number of blockers to eliminate binding of the radioligands to other receptors. This lack of a selective agent prevents their pharmacological characterization. In contrast, the kappa$_3$ receptor has been extensively characterized both in vivo and in binding studies.[2,5-10]

The delta receptor in the NG108-15 cells (DOR-1) was the first opioid receptor cloned[11,12] and provided a tool for the subsequent identification of mu (MOR-1) and kappa (KOR-1) receptor clones.[13-24] When expressed, all the clones demonstrate the anticipated pharmacological selectivity in binding and functional biochemical assays. A fourth clone with high homology to the other opioid receptors has been cloned

Antisense Strategies for the Study of Receptor Mechanisms,
edited by Robert B. Raffa and Frank Porreca. © 1996 R.G. Landes Company.

Table 2.1. Tentative classification of opioid receptor subtypes and their actions

Receptor	Ligand	Analgesia	Other
Mu	Morphine		
	β-Funaltrexamine		
Mu$_1$	Naloxonazine	Supraspinal	Prolactin release
			Feeding
Mu$_2$		Spinal	Respiratory depression
			Gastrointestinal transit
			Feeding
			Guinea pig ileum bioassay
M6G	morphine-6β-glucuronide	Supraspinal and Spinal	
Kappa			
Kappa$_1$	U50,488H	Spinal	Diuresis
	U69,593		Feeding
Kappa$_2$	Bremazocine	Unknown	Unknown
Kappa$_3$	Naloxone Benzoylhydrazone	Supraspinal	Feeding
Delta	Enkephalins		Mouse vas deferens bioassay
Delta$_1$	DPDPE	Supraspinal	
Delta$_2$	[D-Ala^2,Glu^4]Deltorphin	Spinal and Supraspinal	

by our group (KOR-3)[25-27] and others (ORL-1).[25,28-35] However, associating this clone with the established opioid receptors has been difficult. When expressed, it does not show the anticipated affinity or selectivity of any of the pharmacologically defined receptors. Recent work has uncovered a novel peptide which appears to be the endogenous ligand for this receptor.[36,37] Several approaches also suggest that this clone is closely related to the kappa$_3$ receptor,[25-27] as discussed below.

Correlating these cloned opioid receptors with opioid pharmacology is not simple. First, it is important to express the clones and characterize them in both binding and functional assays, such as cyclase. These assays can often establish the selectivity of the receptor for various ligands and can provide an indication of its general class. However, this does not address whether this specific clone is responsible for a specified pharmacological action. Examining the pharmacological significance of the clones can be effectively approached through several strategies. Knocking out the gene of interest with the total loss of the protein is the most definitive approach, but it does have some disadvantages. First, the absence of the protein during development raises the possibility that its functions may be taken over by other proteins. The loss of the protein in question may also prove lethal, making its evaluation impossible. Finally, generating knockout animals requires extensive resources and time, greatly limiting the number of proteins which can be targeted.

Antisense approaches offer an alternative approach.[38,39] In contrast to transgenic mice, antisense strategies can be performed in normal mice. Numerous proteins can be targeted relatively easily and large numbers of animals can be tested. One of the major advantages of this approach is the selectivity of the targeting. Changing the sequence of as few as 4 bases of the 20 comprising most antisense oligonucleotides can eliminate activity.[38] However, these approaches also have some drawbacks.[40,41] Stability of the antisense probes can be a difficult problem, as can getting the probe to the target tissue. Studies in the brain avoid many of these problems. Administering probes directly into brain tissue or into the ventricular or intrathecal cerebrospinal fluid is relatively simple, although diffusion into deeper structures removed from either the surface or the ventricular system may be limited.[42] Furthermore, the probes are far more stable when administered into the central nervous system.[43] Indeed, it is not necessary to use stabilized derivatives such as the phosphorothioates, which is fortunate since they appear to be more toxic than standard deoxynucleotide probes.[41,44] Nonspecific effects also can prove to be problematic and special care must be taken to include extensive controls in all studies.[38,40,41]

Antisense approaches may not be useful for all proteins. The mechanism(s) responsible for the activity of these agents has not been fully worked out.[40,41] Many investigators believe that the probes complex with the mRNA and destabilize it, inducing its downregulation with a resulting decrease in protein expression. While modest decreases can be achieved, it is difficult to completely eliminate the expression of the protein. Thus, the success of an antisense strategy may hinge upon the ability to see a pharmacological or physiological effect with only a partial downregulation of the protein. Finally, care must be used when designing the studies. Although administration of an antisense probe may markedly decrease the synthesis of new protein, time must be given for the preexisting protein to be cycled away.

EFFECTS OF ANTISENSE OLIGODEOXYNUCLEOTIDES ON DOR-1 IN NG108-15 CELLS AND IN VIVO

When developing a new antisense strategy, it is important to demonstrate that the antisense probes reach their target cells, are taken up and downregulate the mRNA and its encoded protein. Our first studies focused upon the DOR-1 clone, which encodes a delta opioid receptor.[45] The receptor had initially been cloned from the NG108-15 cell line and the initial studies examined the actions of antisense oligodeoxynucleotides in this system. We initially chose an antisense oligodeoxynucleotide based upon the first exon of the DOR-1 clone.

When added to cells in the absence of serum, a small fraction of the oligodeoxynucleotide was rapidly taken up into the cells and, once in the cells, was stable for at least 72 hours.[46]

Exposure of the cells to the oligodeoxynucleotides in serum-free media downregulated ^{3}H-DPDPE binding by approximately 40%.[45] All the antisense probes were equally effective regardless of where along the mRNA they were targeted. Thus, antisense oligodeoxynucleotides aimed against exons 1, 2 or 3 all lowered binding equally well (Fig. 2.1). We then explored the actions of the antisense against the first exon in greater detail.[46] In addition to the loss of binding, treatment of the cells with this probe decreased the levels of delta receptor determined by Western analysis. Although suggestive, measurements of mRNA levels were not as clear. Northern analysis of the DOR-1 mRNA reveals multiple transcripts, with the largest corresponding in size to the single transcript seen in spinal cord. Treatment with the antisense lowered the levels of the largest mRNA transcript, but not the smaller ones.

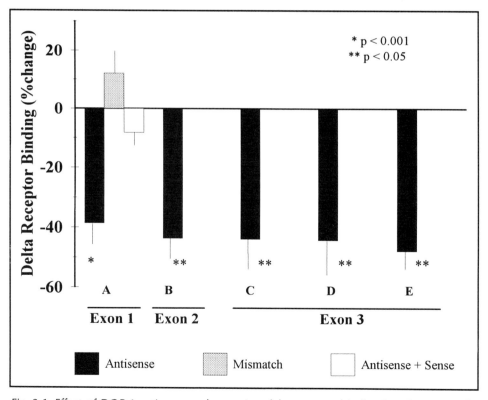

Fig. 2.1. Effect of DOR-1 antisense probes against delta receptor binding in NG108-15 cells. NG108-15 cells were treated with a an antisense oligodeoxynucleotide against the indicated exon of DOR-1 and ^{3}H-DPDPE binding determined. In addition, we examined a mismatch probe against exon 1 and an additional control in which the sense strand was added to the antisense probe to complex it and render it inactive. From the literature.[45]

EFFECT OF ANTISENSE OLIGODEOXYNUCLEOTIDES AGAINST DOR-1 AND KOR-1 IN VIVO

Injection of oligodeoxynucleotides into the central nervous system results in significant uptake into cells of the spinal cord.[42,46] As in the tissue culture system, the oligodeoxynucleotides are stable within the spinal cord for at least 72 hours. Repeated intrathecal treatment with the antisense probe against exon 1 on Days 1, 3 and 5 lowered delta receptor binding in the spinal cord by 30% on Day 6,[45] with a similar reduction in the DOR-1 mRNA levels.[46] The same treatment paradigm significantly lowered the analgesic actions of intrathecal DPDPE (Fig. 2.2) and [D-Ala^2,Glu^4]deltorphin without affecting the actions of mu or kappa$_1$ analgesics. A similar loss of analgesic sensitivity was observed with antisense probes against exons 2 and 3. In these studies, mismatch controls were inactive. Furthermore, analgesic sensitiv-

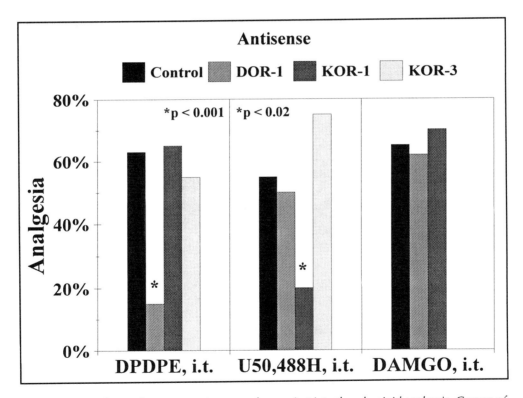

Fig. 2.2. Effects of opioid receptor antisense probes against intrathecal opioid analgesia. Groups of mice were treated intrathecally with antisense probes against a delta (DOR-1), kappa$_1$ (KOR-1) or kappa$_3$-related (KOR-3) clone and analgesia assessed with the delta drug DPDPE, the kappa$_1$ agent U50,488H or the mu drug DAMGO. From the literature.[26,27,45,51]

ity returned slowly over a period of several days, consistent with prior studies examining the turnover of the receptor. Other groups subsequently have utilized similar approaches and also confirm the involvement of the DOR-1 in delta analgesia,[47-49] although there is a suggestion of delta subtypes from these antisense studies.[50]

Antisense studies in mice[51] and in rats[52] also have confirmed the importance of the KOR-1 clone in kappa$_1$ analgesia (Fig. 2.2). Again, the actions of the antisense probes were restricted to kappa drugs; mu and delta analgesia was unaffected. Mismatch controls also were without activity.

ANTISENSE MAPPING MOR-1

The ability to downregulate delta receptor binding in NG108-15 cells by targeting any of the three exons suggested that antisense strategies could be used to explore the role of individual exons in pharmacological actions and the possibility of alternative splicing. We first examined the effects of an antisense against the 5'-untranslated region of MOR-1 microinjected into the periaqueductal gray of the rat.[53] This injection paradigm dramatically lowered morphine analgesia. Later studies indicated that it has little effect upon the analgesic actions of the delta ligand [D-Ala^2,Glu^4]deltorphin, confirming its selectivity. Similar results have been obtained by others.[54]

We then explored a series of antisense oligodeoxynucleotides targeting exons 1, 2, 3 and 4 of MOR-1 in the mouse (Fig 2.3).[39,53] Three separate probes against exon 1 all blocked supraspinal morphine analgesia, as did the probe against the coding region of exon 4. However, none of the three targeting exon 2 and only one of three directed against exon 3 were active. Pharmacological studies have suggested that supraspinal morphine analgesia is mediated through the mu$_1$ receptor subtype while the inhibition of gastrointestinal transit by supraspinal morphine is elicited through mu$_2$ receptors. When we examined the activity of the antisense probes in the GI transit assay, only the probe based upon the coding region of exon 4 was active. Thus, the profile of the antisense probes against supraspinal morphine analgesia and the inhibition of GI transit differed. Spinal morphine analgesia also has been classified as a mu$_2$ action. Its sensitivity profile was the same as supraspinal GI transit effects.

Morphine-6β-glucuronide (M6G) is a potent morphine metabolite.[55-59] When we examined the sensitivity of M6G analgesia to the various antisense oligodeoxynucleotides we observed a very different profile.[39,53] The three probes against exon 1 and the one against the coding region of exon 4 all were inactive. In contrast, the six probes against exons 2 and 3 all effectively blocked M6G analgesia, suggesting the existence of a previously unrecognized M6G receptor which appears to be a splice variant of MOR-1. Together, these differing sensitivity profiles against the various antisense probes raises the possi-

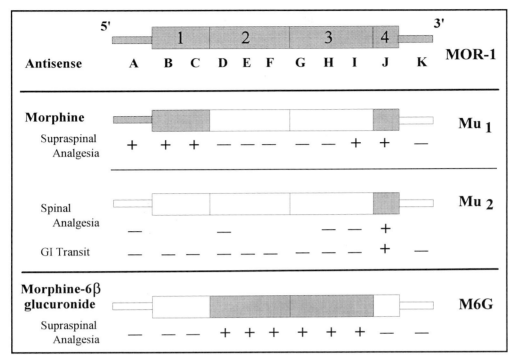

Fig. 2.3. Antisense mapping MOR-1 on morphine and morphine-6β-glucuronide actions. Antisense probes targeting the indicated exons of MOR-1 were tested in groups of mice against the stated actions with either morphine or morphine-6β-glucuronide. Results are from the literature.[53,62]

bility that the mu receptor subtypes and the putative M6G receptor might represent splice variants of the MOR-1 clone. However, until these potential variants have been cloned, this remains only a hypothetical explanation of these results.

ANTISENSE MAPPING KOR-3

Soon after the cloning of the DOR-1 receptor, many groups embarked upon studies looking for homologous clones. Like other groups, we identified a novel cDNA using RT-PCR based upon degenerate primers of the DOR-1 clone.[25-27] Traditional approaches involve the subsequent cloning of the full length cDNA, followed by its characterization after expression. Having had success with the antisense approach against DOR-1, KOR-1 and MOR-1, we designed an antisense probe targeting the cDNA fragment and examined it against the analgesic actions of a series of opioids before cloning the full length cDNA (Fig. 2.4). Treating animals with the antisense oligodeoxynucleotide had no effect against mu, delta or kappa$_1$ analgesia. However, kappa$_3$ analgesia was effectively blocked, suggesting that the sequence we had

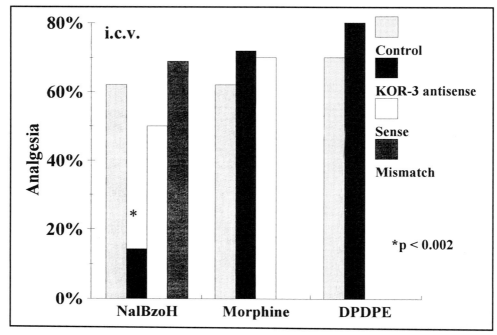

Fig. 2.4. Effects of a KOR-3 antisense probe on opioid analgesia. An antisense probe based upon a partial cDNA fragment from an RT-PCR reaction was tested against mu (morphine), delta (DPDPE) and kappa$_3$ (NalBzoH) opioid analgesia. The control group received vehicle and an additional group received mismatch. From the literature.[26,27]

obtained was contained within the kappa$_3$ receptor. We then cloned the full length cDNA and examined it in binding and functional assays. Our clone, KOR-3, is the murine version of clones reported by other groups (ORL-1).[25,28-35] In addition to the antisense studies, we also demonstrated that the receptor encoded by the KOR-3 clone was recognized by a monoclonal antibody[27] directed against the kappa$_3$ receptor present in a human neuroblastoma cell line.[60] While these studies suggested that the receptor fell within the opioid receptor family, the expressed clone did not demonstrate the selectivity and affinities in either binding or functional assays anticipated from prior studies with the kappa$_3$ receptor in brain and cell lines. The identification of a novel peptide with high affinity for this clone has opened new areas of investigation.[36,37]

To explore the relationship between KOR-3 and the kappa$_3$ receptor, we performed antisense mapping of the KOR-3 clone (Fig. 2.5).[27] The coding region of the KOR-3 gene[61] is comprised of three exons, with splice sites within the first transmembrane region and in the second extracellular loop. Six antisense oligodeoxynucleotides targeting the second and third coding exons all effectively lowered kappa$_3$ analgesia. However, only one of an additional six probes directed against the

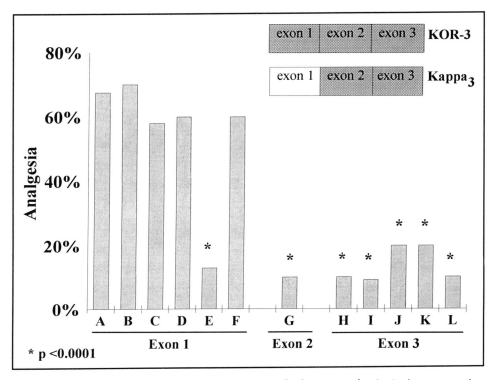

Fig. 2.5. Antisense mapping KOR-3 on naloxone benzoylhydrazone analgesia. Antisense mapping of KOR-3 using antisense probes targeting the indicated exon. The insert illustrates the possible relationship between the KOR-3 clone and the kappa$_3$ receptor, in which they share exons 2 and 3 but differ in their first exons. From the literature.[27]

first coding exon was active. This strong demarcation between the exon encoding the NH$_2$-terminus and the two others further downstream raises the possibility that the kappa$_3$ receptor shares two exons with the KOR-3 clone, with the two receptors differing at the first exon. Unfortunately, the meaning of inactivity of the probes against the first coding exon is not entirely clear. While it probably reflects the absence of these sequences in the kappa$_3$ receptor clone, it might be due to secondary structures in the mRNA or other technical factors. In contrast, virtually all the antisense probes in the MOR-1 study were active in at least one assay.

IMPLICATIONS FOR THE FUTURE

Antisense studies have enormous potential in exploring the functional roles of selected proteins. Its major advantage is its extraordinary selectivity, utility in normal animals and the ability to target large number of proteins relatively easily. Within the opioid field, antisense strategies have a selectivity for individual receptors which far exceed those of any antagonist. The ability to examine individual exons offers the additional possibility of looking at splice variants in a functional context. The central nervous system, in general, may be particularly suited to these types of studies. Antisense can be administered directly into the region of interest and the oligonucleotides are far more stable than after systemic administration. The opioid system has a number of advantages in these approaches. Foremost is the localization of the receptors mediating analgesia, which are present in the periventricular regions of the brainstem and in the dorsal horn of the spinal cord. These areas are easily reached by diffusion of the oligodeoxynucleotides. Despite these advantages, a number of issues must be considered when using antisense strategies. The mechanism through which antisense approaches produce their actions is still unclear. The antisense oligonucleotides might act either through a disruption of translation or by destabilizing mRNA. However, nonspecific actions and toxicity also have been reported.[41,44] These problems can be minimized by using the lowest doses needed to see an effect, but extreme care must be exercised to avoid being misled. It is often quite difficult to downregulate the protein totally, particularly when using low antisense doses. Thus, the assay of protein function should be able to detect loss of only 30% of the protein. Finally, these approaches also may not prove useful for all proteins.

ACKNOWLEDGMENTS

We thank Dr. Jerome Posner for his assistance with these studies. The work presented was supported by research grants (DA02615, DA07242 and DA6241) from the National Institute on Drug Abuse. GWP is supported by a Research Scientist Award (DA00220) from the National Institute on Drug Abuse.

REFERENCES

1. Pasternak GW. Pharmacological mechanisms of opioid analgesics. Clin Neuropharmacol 1993; 16:1-18.
2. Clark JA, Liu L, Price M et al: Kappa opiate receptor multiplicity: evidence for two U50,488-sensitive $kappa_1$ subtypes and a novel $kappa_3$ subtype. J Pharmacol Exp Ther 1989; 251:461-468.
3. Rothman RB, Bykov V, DeCosta BR et al: Interaction of endogenous opioid peptides and other drugs with four kappa opioid binding sites in guinea pig brain. Peptides 1990; 11:311-317.
4. Zukin RS, Eghbali M, Olive D et al. Characterization and visualization of of rat and guinea pig brain kappa opioid receptors: evidence for $kappa_1$ and $kappa_2$ opioid receptors. Proc Natl Acad Sci USA 1988; 85:4061-4065.
5. Paul D, Levison JA, Howard DH et al. Naloxone benzoylhydrazone (NalBzoH) analgesia. J Pharmacol Exp Ther 1990; 255:769-774.
6. Price M, Gistrak MA, Itzhak Y et al. Receptor binding of ^{3}H-naloxone benzoylhydrazone: a reversible kappa and slowly dissociable μ opiate. Mol Pharmacol 1989; 35:67-74.

7. Gistrak MA, Paul D, Hahn EF et al. Pharmacological actions of a novel mixed opiate agonist/antagonist, naloxone benzoylhydrazone. J Pharmacol Exp Ther 1990; 251:469-476.
8. Standifer KM, Cheng J, Brooks AI et al. Biochemical and pharmacological characterization of mu, delta and kappa$_3$ opioid receptors expressed in BE(2)-C neuroblastoma cells. J Pharmacol Exp Ther 1994; 270:1246-1255.
9. Cheng J, Standifer KM, Tublin PR et al. Demonstration of kappa$_3$-opioid receptors in the SH-SY5Y human neuroblastoma cell line. J Neurochem 1995; 65:170-175.
10. Paul D, Pick CG, Tive LA et al. Pharmacological characterization of nalorphine, a kappa$_3$ analgesic. J Pharmacol Exp Ther 1991; 257:1-7.
11. Evans CJ, Keith DF, Morrison H et al. Cloning of the delta opioid receptor by functional expression. Science 1992; 258:1952-1955.
12. Kieffer BL, Befort K, Gaveriaux-Ruff C et al. The δ-opioid receptor: isolation of a cDNA by expression cloning and pharmacological characterization. Proc Natl Acad Sci USA 1992; 89:12048-12052.
13. Chen Y, Mestek A, Liu J et al. Molecular cloning and functional expression of a μ-opioid receptor from rat brain. Mol Pharmacol 1993; 44:8-12.
14. Minami M, Toya T, Katao Y et al. Cloning and expression of a cDNA for the rat kappa-opioid receptor. FEBS Lett 1993; 329:291-295.
15. Eppler CM, Hulmes JD, Wang J-B et al. Purification and partial amino acid sequence of a μ opioid receptor from rat brain. J Biol Chem 1993; 268:26447-26451.
16. Reisine T, Bell GI: Molecular biology of opioid receptors. Trends Neurosci 1993; 16:506-510.
17. Wang JB, Johnson PS, Persico AM et al. Human μ opiate receptor: cDNA and genomic clones, pharmacologic characterization and chromosomal assignment. FEBS Lett 1994; 338:217-222.
18. Raynor K, Kong H, Chen Y et al. Pharmacological characterization of the cloned kappa-, δ-, and μ-opioid receptors. Mol Pharmacol 1994; 45:330-334.
19. Yasuda K, Raynor K, Kong H et al. Cloning and functional comparison of kappa and δ opioid receptors from mouse brain. Proc Natl Acad Sci USA 1993; 90:6736-6740.
20. Kozak CA, Filie J, Adamson MC et al. Murine chromosomal location of the μ and kappa opioid receptor genes. Genomics 1994; 21:659-661.
21. Min BH, Augustin LB, Felsheim RF et al. Genomic structure and analysis of promoter sequence of a mouse μ opioid receptor gene. Proc Natl Acad Sci USA 1994; 91:9081-9085.
22. Knapp RJ, Malatynska E, Fang L et al. Identification of a human delta opioid receptor: cloning and expression. Life Sci 1994; 54:PL463-PL469.
23. Bare LA, Mansson E, Yang D: Expression of two variants of the human μ opioid receptor mRNA in SK-N-SH cells and human brain. FEBS Lett 1994; 354:213-216.
24. Zimprich A, Bacher B, Höllt V: Cloning and expression of an isoform of the rmu-opioid receptor (rmuOR1B). Regul Pept 1994; 54:347-348.

25. Uhl GR, Childers S, Pasternak GW: An opiate-receptor gene family reunion. Trends Neurosci 1994; 17:89-93.
26. Pan YX, Cheng J, Xu J et al. Cloning, expression and classification of a kappa$_3$-related opioid receptor using antisense oligodeoxynucleotides. Regul Pept 1994; 54:217-218.
27. Pan Y-X, Cheng J, Xu J et al. Cloning and functional characterization through antisense mapping of a kappa$_3$-related opoioid receptor. Mol Pharmacol 1995; 47:1180-1188.
28. Chen Y, Fan Y, Liu J et al. Molecular cloning, tissue distribution and chromosomal localization of a novel member of the opioid receptor gene family. FEBS Lett 1994; 347:279-283.
29. Keith Jr D, Maung T, Anton B et al. Isolation of cDNA clones homologous to opioid receptors. Regul Pept 1994; 54:143-144.
30. Wick MJ, Minnerath SR, Lin X et al. Isolation of a novel cDNA encoding a putative membrane receptor with high homology to the cloned μ, δ, and kappa opioid receptors. Mol Brain Res 1994; 27:37-44.
31. Wang JB, Johnson PS, Imai Y et al. cDNA cloning of an orphan opiate receptor gene family member and its splice variant. FEBS Lett 1994; 348:75-79.
32. Mollereau C, Parmentier M, Mailleux P et al. ORL-1, a novel member of the opioid family: cloning, functional expression and localization. FEBS Lett 1994; 341:33-38.
33. Bunzow JR, Saez C, Mortrud M et al. Molecular cloning and tissue distribution of a putative member of the rat opioid receptor gene family that is not a μ, δ or kappa opioid receptor type. FEBS Lett 1994; 347:284-288.
34. Fukuda K, Kato S, Mori K et al. cDNA cloning and regional distribution of a novel member of the opioid receptor family. FEBS Lett 1994; 343:42-46.
35. Lachowicz JE, Shen Y, Monsma Jr FJ et al. Molecular cloning of a novel G protein-coupled receptor related to the opiate receptor family. J Neurochem 1995; 64:34-40.
36. Meunier JC, Mollereau C, Toll L et al. Isolation and structure of the endogenous agonist of the opioid receptor-like ORL$_1$ receptor. Nature 1995; 377:532-535.
37. Reinscheid RK, Nothacker HP, Bourson A et al. Orphanin FQ: a neuropeptide that activates an opioidlike G protein-coupled receptor. Science 1995; 270:792-794.
38. Wahlestedt C: Antisense oligonucleotide strategies in neuropharmacology. Trends Pharmacol Sci 1994; 15:42-46.
39. Pasternak GW, Standifer KM. Mapping of opioid receptors using antisense oligodeoxynucleotides: correlating their molecular biology and pharmacology. Trends Pharmacol Sci 1995; 16:344-350.
40. Wagner RW. Gene inhibition using antisense oligodeoxynucleotides. Nature 1994; 372:333-335.
41. Crooke ST. Progress toward oligonucleotide therapeutics: pharmacodynamic properties. FASEB J 1993; 7:533-539.

42. Yee F, Ericson H, Reis DJ et al. Cellular uptake of intracerebroventricularly administered biotin- or digoxigenin-labeled antisense oligodeoxynucleotides in the rat. Cell Mol Neurobiol 1994; 14:475-486.
43. Whitesell L, Gelelowitz D, Chavany C et al. Stability, clearance and disposition of intraventricularly administered oligodeoxynucleotides: implications for therapeutic application with the central nervous system. Proc Natl Acad Sci USA 1993; 90:4665-4669.
44. Chiasson BJ, Armstrong JN, Hooper ML et al. The application of antisense oligonucleotide technology to the brain: some pitfalls. Cell and Mol Neurobiol 1994; 14:507-521.
45. Standifer KM, Chien C-C, Wahlestedt C et al. Selective loss of δ opioid analgesia and binding by antisense oligodeosynucleotides to a δ opioid receptor. Neuron 1994; 12:805-810.
46. Standifer KM, Jenab S, Su W et al. Antisense oligodeoxynucleotides to the cloned δ receptor, DOR-1: Uptake, stability and regulation of gene expression. J Neurochem 1995; 65:1981-1987.
47. Bilsky EJ, Bernstein RN, Pasternak GW et al. Selective inhibition of [D-Ala^2,Glu^4]deltorphin antinociception by supraspinal, but not spinal, administration of an antisense oligodeoxynucleotide to an opioid delta receptor. Life Sci 1994; 55:37-43.
48. Lai J, Bilsky EJ, Porreca F. Treatment with antisense oligodeoxynucleotide to a conserved sequence of opioid receptors inhibits antinociceptive effects of delta subtype selective ligands. J Recept Res 1995; 15:643-650.
49. Tseng LF, Collins KA, Kampine JP: Antisense oligodeoxynucleotide to a δ-opioid receptor selectively blocks the spinal antinociception induced by δ-, but not μ- or kappa-opioid receptor agonists in the mouse. Eur J Pharmacol 1994; 258:R1-R3.
50. Lai J, Bilsky EJ, Rothman RB et al. Treatment with antisense oligodeoxynucleotide to the opioid δ receptor selectively inhibits δ$_2$-agonist antinociception. Neuroreport 1994; 5:1049-1052.
51. Chien C-C, Brown G, Pan Y-X et al. Blockade of U50,488H analgesia by antisense oligodeoxynucleotides to a kappa-opioid receptor. Eur J Pharmacol 1994; 253:R7-R8.
52. Adams JU, Chen X, Deriel JK et al. Intracerebroventricular treatment with an antisense oligodeoxynucleotide to kappa-opioid receptors inhibited kappa-agonist-induced analgesia in rats. Brain Res 1994; 667:129-132.
53. Rossi GC, Pan Y-X, Brown GP et al. Antisense mapping the MOR-1 opioid receptor: evidence for alternative splicing and a novel morphine-6β-glucuronide receptor. FEBS Lett 1995; 369:192-196.
54. Chen XH, Adams JU, Geller EB et al. An antisense oligodeoxynucleotide to μ-opioid receptors inhibits μ-opioid receptor agonist-induced analgesia in rats. Eur J Pharmacol 1995; 275:105-108.
55. Paul D, Standifer KM, Inturrisi CE et al. Pharmacological characterization of morphine-6β-gluruconide, a very potent morphine metabolite. J Pharmacol Exp Ther 1989; 251:477-483.
56. Pasternak GW, Bodnar RJ, Clark JA et al. Morphine-6-glucuronide, a potent mu agonist. Life Sci 1987; 41:2845-2849.

57. Shimomura K, Kamata O, Ueki S et al. Analgesic effect of morphine glucuronides. Tohoku J Exp Med 1971; 105:45-52.
58. Abbott FV, Palmour RM: Morphine-6-glucuronide: analgesic effects and receptor binding profile in rats. Life Sci 1988; 43:1685-1695.
59. Sullivan AF, McQuay HJ, Bailey D et al. The spinal antinociceptive actions of morphine metabolites, morphine-6β-glucurnoide and normorphine in the rat. Brain Res 1989; 482:219-224.
60. Brooks AI, Standifer KM, Rossi GC et al. Characterizing kappa$_3$ opioid receptors with a selective monoclonal antibody. Synapse 1995; in press.
61. Pan Y-X, Xu J, Pasternak GW: Structure and characterization of the gene encoding a mouse kappa$_3$-related opioid receptor. Gene 1995; in press.
62. Rossi GC, Standifer KM, Pasternak GW: Differential blockade of morphine and morphine-6β-glucuronide analgesia by antisense oligodeoxynucleotides directed against MOR-1 and G-protein α subunits in rats. Neurosci Lett 1995; 198:99-102.

CHAPTER 3

ANTISENSE TARGETING OF DELTA OPIOID RECEPTORS

Josephine Lai and Frank Porreca

ANTISENSE OLIGODEOXYNUCLEOTIDE MEDIATED GENE TARGETING

Gene targeting by antisense oligodeoxynucleotides (ODN) is based on the ability of short, synthetic, single-stranded DNA to interdict gene expression in a sequence specific manner. ODNs are thought to interfere with the biosynthesis of a particular gene product by binding to complementary sequences in the target gene or its messenger RNA, thereby disrupting the normal transcription or translation of that gene.[1] By virtue of the high affinity and specificity of the ODN for their target sequences, the resultant "knock-down" of the target protein could be orders of magnitude greater in specificity than conventional antagonists, and is directly correlated with the known structural characteristics of that protein. These potential advantages of antisense ODN form the rationale for developing ODN as therapeutic agents, or so-called "information drugs" because of their design, particularly as antiviral and anticancer drugs.[2] Numerous reports in the literature attest to the effectiveness of targeting viral proteins or proto-oncogenes both in vitro and in animal models. In more recent years, this technology has also attracted growing interest within the neuroscience community in using antisense ODN mediated "knock-down" as a means to manipulate specific proteins in the central nervous system (for recent reviews, see refs. 3,4). Besides the inherent specificity of antisense ODN for their molecular target, applying antisense ODN to in vivo targets has the potential for pinpointing the molecular basis for a defined physiological function. Targeting may also specify gross anatomically defined sites to provide neuroanatomical relevance. Antisense strategy

---

Antisense Strategies for the Study of Receptor Mechanisms,
edited by Robert B. Raffa and Frank Porreca. © 1996 R.G. Landes Company.

is thus an innovative approach which may allow us to examine directly the function of structurally identified proteins in a living animal.

It must be emphasized that as with any novel experimental approach, the use of antisense ODN to target specific receptors in the CNS must be carefully considered to ensure the specificity, reproducibility and validity of such approach. There are several technical issues that are often raised about antisense targeting, which are of critical significance for developing antisense ODN as therapeutic agents, as well as for using antisense ODN solely as an analytical tool. These include the uptake and stability of the antisense ODN; the specificity of the antisense ODN for the target protein and their potential nonspecific or toxic effects; and appropriate controls for these experiments. We will consider each of these issues in the following discussion of our studies in which antisense ODN were used to target the cloned δ opioid receptors in mice.

PHARMACOLOGICAL EVIDENCE FOR SUBTYPES OF THE δ OPIOID RECEPTOR

There are three pharmacologically distinct opioid receptors, termed μ, δ and κ. Heterogeneity of these three receptor types has also been proposed. In the case of the δ receptors, the strongest pharmacological evidence for the existence of opioid δ receptors has come from experiments in vivo. The first line of evidence is based on the ability of δ selective antagonists to differentially inhibit the supraspinal antinociceptive effect of δ receptor agonists. In these studies, two highly δ selective peptides were used, namely [D-Pen^2, D-Pen^5]enkephalin (DPDPE) and [D-Ala^2, Glu^4]deltorphin, which produced antinociception in the mouse warm-water tail-flick test via supraspinal δ receptors. Their effects could both be antagonized by the δ selective antagonist, ICI 174,864, but not by the μ selective antagonist, β-FNA.[5,6] The antinociceptive actions of DPDPE could be blocked in a time and dose related manner by the nonequilibrium antagonist [D-Ala^2, Leu^5, Cys^6]enkephalin (DALCE).[7] However, DALCE pretreatment at the same doses and times did not antagonize the antinociceptive action of [D-Ala^2, Glu^4]deltorphin.[7] Concurrently, another δ selective antagonist, 5'-naltrindole isothiocyanate (5'-NTII)[8], was shown to inhibit the antinociceptive action of [D-Ala^2, Glu^4]deltorphin in a time and dose dependent manner, but did not affect the antinociception of DPDPE.[7] The ability of DALCE and 5'-NTII to two-way differentially antagonize the antinociceptive effects of DPDPE and [D-Ala^2, Glu^4]deltorphin, respectively, suggests that the action of the two δ selective agonists may be mediated by distinct δ receptor subtypes.

The second line of evidence that implicates δ receptor subtypes came from the induction of tolerance to DPDPE and [D-Ala^2, Glu^4]deltorphin upon chronic supraspinal administration in mice.

Repeated intracerebroventricular (i.c.v.) injection of DPDPE led to a 5-fold rightward displacement of the dose response for the peptide.[9] In these mice, the dose response for [D-Ala^2, Glu^4]deltorphin was the same as control, suggesting a lack of cross-tolerance to [D-Ala^2, Glu^4]deltorphin in these DPDPE tolerant mice. Similarly, tolerance could be developed to [D-Ala^2, Glu^4]deltorphin through chronic i.c.v. administration, shown by a 37-fold rightward shift in the dose-response for this peptide. The antinociceptive effect of DPDPE was unaffected in these mice. In addition, mice that had developed tolerance to the μ agonist, DAMGO, did not show cross-tolerance to either DPDPE or [D-Ala^2, Glu^4]deltorphin. Hence, based on the two-way differential antagonism of DPDPE and [D-Ala^2, Glu^4]deltorphin, and their two-way lack of antinociceptive cross-tolerance, it was concluded there were at least two distinct pharmacological subtypes of δ receptors. The term δ$_1$ is used to describe the site acted upon by DPDPE and antagonized by DALCE, and δ$_2$ describes that which is acted upon by [D-Ala^2, Glu^4]deltorphin and antagonized by 5'-NTII. Both receptor subtypes are insensitive to β-FNA and naloxonazine (μ), but are sensitive to ICI 174,864.[10] It should be noted that DPDPE is not completely selective for the δ$_1$ subtype, however, its direct antinociceptive effects appear to be mediated almost entirely at this site.[11]

Subtypes of δ opioid receptors have also been demonstrated at the spinal level in the rat. Intrathecal administration of naltriben (NTB) blocked the antinociceptive actions of [D-Ala^2, Glu^4]deltorphin but not that of DPDPE or μ receptor agonists.[12] NTB is thus considered to act preferentially at the δ$_2$ receptor site. These findings were in agreement with conclusions from other studies using site directed administration of selective agonists and antagonists in the rat.[13,14] With regard to the selectivity of DPDPE, as mentioned above, it is clear from other investigations that DPDPE is not completely selective for the proposed δ$_1$ subtype.[11] In this regard, DPDPE has been shown to elicit the characteristic modulatory action of δ agonists on μ-mediated antinociception via a δ$_2$ receptor,[15] and when δ$_1$ receptors are blocked with DALCE, to antagonize the antinociceptive actions of [D-Ala^2,Glu^4]deltorphin.[11]

Despite strong evidence in the in vivo studies that suggest the existence of multiple δ receptors, in vitro analysis of brain tissues using radioligand binding has not established a clear identity of the pharmacological δ$_1$ and δ$_2$ receptor subtypes through differential binding affinity of δ selective ligands to these sites. Molecular cloning has identified one gene whose product conforms to the pharmacologically defined δ type,[16,17] and attempts in identifying structurally related genes have so far not yielded any other gene products that may be clearly defined as a δ subtype. On the other hand, a recent communication suggests that a receptor from immune cells may pharmacologically conform to that of a δ receptor (Burt Sharp, communication) and which shares little structural homology with the previously cloned gene. The nature of the immune system derived opioid receptor remains to be verified. Because the pharmacologically defined δ receptor subtypes are based on in vivo rather than in vitro studies, the hypothesis of δ subtypes could not be adequately tested on the cloned δ receptor simply through in vitro expression and characterization of this receptor. We thus sought to test the δ subtype hypothesis using the antisense strategy, in which antisense ODN were designed based on the structure of the cloned δ receptor and used to target the δ selective ligand mediated antinociception in mice. The basic premise is that antisense ODN that specifically target the cloned δ receptor should block the antinociception of drugs that are mediated by these receptors. Thus, by using δ selective agonists that have been used previously to define the δ_1 and δ_2 sites, the effect of the antisense ODN on the function of these ligands may substantiate δ receptor heterogeneity, and further, provide a correlation between the pharmacologically defined δ receptors and the structure of the receptor as defined by the cloned gene.

SEQUENCE-SPECIFIC BLOCKADE OF δ OPIOID RECEPTOR ANTINOCICEPTION IN VIVO

CHOICE OF SEQUENCE FOR THE ANTISENSE ODN

As for any drug design, the primary consideration for antisense ODN is their specificity for the gene target, which is governed by the sequence complementarity between the antisense ODN and its target. Antisense ODN sequences are often derived from within the cDNA for the target protein, thus most antisense ODN target the coding sequences of a gene, or the immediate flanking untranslated regions present in the transcripts. It has been estimated that short ODN between 11 and 15 bases in length should bind with high affinity and selectivity to a single RNA species in the cell (for review, see ref. 18). The formation of the RNA/DNA hybrid between the transcript and the ODN may interfere with protein biosynthesis by two established mechanisms: translation arrest and/or degradation of the target transcript by RNase H, which recognizes the RNA/DNA hybrid as substrate. The 5' end of the coding region is generally considered to be a suitable target sequence in the design of antisense ODN because of its proximity to the initiation of translation. RNase H activity, on the other hand, may be induced by the RNA/DNA hybrid formation anywhere within the transcript. Inhibition of protein synthesis by antisense ODN either through translation arrest or activation of RNase H has been well demonstrated in vitro. It is conceivable that antisense ODN derived from unique sequences proximal to the 5' end of the coding region of the target transcript should therefore be highly effective.

In the case of designing antisense ODN specific for the δ receptor, the structural conservation among the cloned δ, μ and κ opioid receptors, or with other structurally related genes (including the somatostatin receptor) has to be considered. The most structurally divergent regions include the 5' end (encoding the amino terminus) and the 3' end (encoding the carboxyl terminus) of the coding region, while regions that encode the putative transmembrane domains of the receptor share substantial sequence homology with that of the other opioid receptors.[19] A tentative target sequence at the 5' end of the coding region (nucleotides 7 to 26) for the δ receptor was therefore assigned to specify the δ receptor. Ideally, this sequence is unique to the δ receptor gene, and other genes which contain homologous sequence should not bind the antisense ODN because of mismatch in the sequence. To test this, the antisense sequence to this was screened against the GenBank database to identify any substantial homology with other known genes. This is a common practice during the design of the antisense even though it is somewhat arbitrary because of the limitation of the database, and because factors other than the sequence may influence the ODN/RNA interaction, including pH, ionic strength, secondary structure of the transcript and the chemical composition of the antisense ODN. In general, studies in which antisense ODN are applied in vivo or in cell cultures, the specificity and the effectiveness of the antisense ODN for their target are determined empirically using carefully defined control ODN as discussed below. This is in part due to the difficulty in predicting the effectiveness of the antisense ODN based on sequence alone, and the technical problem of demonstrating the binding of the antisense ODN to its presumed target transcript in vivo.

STABILITY OF THE ANTISENSE ODN

The stability of ODN has been a major focus in the development of antisense strategy because of the rapid degradation of natural ODN in serum and in various tissues (for review, see ref. 18). The phosphorothioate derivative of ODN has been widely tested because of its enhanced resistance against nuclease activity. However, these molecules also appear to be more toxic when compared with the phosphodiester analogs,[20] and may be due to interaction with cellular proteins.[21] The use of unmodified, antisense ODN appeared to be effective when administered directly into the brain[22] or spinal cord.[23]

UPTAKE OF THE ANTISENSE ODN

Because the target for antisense ODN is intracellular, their translocation into cells which express the target protein is critical for their action. Our initial experiments using fluorescence tagged ODN showed that the neuroblastoma/glioma cell line, NG 108-15, which we use as an in vitro model for δ receptor targeting, accumulated the ODN in a time and concentration dependent manner.[24] These observations are consistent with a number of previous studies in which phosphodiester

ODN[25,26] or phosphorothioate ODN[27,28] were shown to be taken up by a number of different cell lines, and this uptake was thought to be mediated by both adsorptive endocytosis and fluid-phase pinocytosis. Fluorescence was detected in both the cytoplasm and the nucleus, suggesting that the ODN had access to both compartments, similar to observations made with other cell types which were exposed to ODN.[25,27,28] Uptake of ODN has also been observed in vivo through direct i.c.v. injection.[29] Interestingly, the ODN was not taken up equally by all cells, both in vivo as well as in culture, and may be influenced by the metabolic activity of the cells.

CONTROLS

The design and use of appropriate controls for antisense ODN mediated targeting constitutes the most important parameter in these studies. Controls are ODN which sufficiently mimic the chemical composition of the ODN but without the target specificity of the antisense ODN. Thus control ODN are usually of the same length and chemical make-up as the antisense ODN, i.e., same type of nucleotide analog, same GC content, but whose sequence are not specific for the desired target transcripts. These molecules are designed to affirm the sequence specificity of the antisense ODN for the target and to assess the degree of sequence-independent effects of ODN treatment in the functional assays. Conceptually, these control ODN are not equipped to mimic the inadvertent interaction of the antisense ODN with other sequences based on sequence complementarity, because the sequence of the control ODN is different from that of the antisense ODN. In this regard, mismatch ODN, whose sequence is that of the antisense ODN but scrambled at several positions either by substitution or by shuffling two adjacent bases at random, most closely resemble the sequence of the antisense ODN while its affinity and specificity for the target are greatly diminished. The design of the control ODN for the δ specific antisense ODN was based on shuffling the antisense sequence at three paired positions to generate six mismatches out of 20 bases. This mismatch ODN sequence therefore shares 70% identity with that of the antisense ODN. Other design of control ODN include using the complementary sense sequence (target sequence), or "random sequence" that bears no relation to the antisense sequence.

TREATMENT PARADIGM

The treatment paradigm was established empirically through preliminary analysis of time course and dose range of i.c.v. administration of the antisense ODN for the δ receptor into ICR mice, and using the warm-water tail-flick test as the assay for antinociception.[30] It was found that δ antisense ODN inhibited the antinociceptive effect of [D-Ala^2,Glu^4]deltorphin maximally after a 3-day treatment with two bolus injections daily at 12.5 μg/injection of the antisense ODN (phosphodiester form; reconstituted in nuclease free water). Treated animals continued to feed normally and displayed typical behavior. The onset of the biological effect of the antisense ODN presumably depends on the turnover rate of the existing δ receptors in the brain, the rate and extent of ODN accumulation in the tissue.

EFFECT OF δ ANTISENSE ODN ON THE ANTINOCICEPTION OF δ RECEPTOR SELECTIVE AGONISTS

Initial experiments showed that this treatment paradigm of mice with the δ antisense ODN, but not the control ODN, effectively inhibited the antinociception of the δ_2 selective agonist, [D-Ala^2,Glu^4]deltorphin in both the warm-water tail-flick and the hot plate test.[31] On the other hand, the antinociceptive effect of the δ_1 selective agonist, DPDPE, was not altered by this antisense ODN, or the control ODN. These findings suggest that the observed effects were sequence specific and not due to the ODN treatment on behavior or nociceptive sensitivity. The selective inhibition of the antinociceptive effect of a δ_2, but not of a δ_1 agonist, suggests that the cloned δ receptor may be the same as that pharmacologically classified as δ_2, thus providing an additional line of evidence to support the existence of multiple opioid δ receptors. This line of study was also used to examine the δ receptors in the spinal cord through intrathecal injection of the ODN.[23,32] It was found that the control ODN did not alter the baseline latencies compared with the control group or the antinociceptive response to opioid agonists DPDPE (δ_1), [D-Ala^2,Glu^4]deltorphin (δ_2), DAMGO (μ) or U69,593 (κ). However, the δ antisense ODN treated animals showed a significant reduction in the antinociceptive response to [D-Ala^2,Glu^4]deltorphin as well as DPDPE, while the antinociceptive response to DAMGO or U69,593 was not affected. The finding that both proposed δ_1 and δ_2 agonists were sensitive to the δ antisense ODN treatment at the spinal level would appear to contradict the suggestion that these agonists exert their antinociceptive effects via distinct subtypes of the δ receptor. However, it has been previously shown, based on the effect of DALCE and 5'-NTII at the spinal level, that a single type of δ opioid receptors predominates in the spinal cord.[10] As noted above, DPDPE does not act exclusively at the δ_1 site. Thus, these data are consistent with previous studies which suggest that the 5'-NTII sensitive δ_2 sites at the spinal level mediate both the antinociceptive response of [D-Ala^2,Glu^4]deltorphin and DPDPE.

Further analysis on the i.c.v. administration of the δ antisense ODN in mice showed that the dose-dependent antinociceptive effect of [D-Ala^2,Glu^4]deltorphin was shifted significantly to the right (7-fold), while that of DPDPE was not affected.[30] Furthermore, there was a concomitant reduction in the density of the δ receptors based on [^{3}H]naltrindole binding of whole brain membranes. The effect of the δ antisense ODN on the antinociceptive response to

[D-Ala²,Glu⁴]deltorphin, as well as the δ receptor density, was fully reversible such that upon cessation of ODN treatment, the antinociceptive response to [D-Ala²,Glu⁴]deltorphin and the level of [³H]naltrindole binding returned to control levels within three days. The δ antisense ODN treatment did not affect the antinociceptive effect of either κ or μ selective agonists, further substantiating the specificity of the ODN for the δ receptor.

An important control experiment in support of the selectivity of the δ antisense ODN for δ_2 mediated response is to show that, in fact, that DPDPE could be blocked by an antisense ODN that may target its site of action, or the putative δ_1 site. However, because the molecular basis for this site has not been established, specific targeting of this site by antisense strategy is not feasible. We thus made the assumption that an antisense ODN that targets sequences conserved among the cloned opioid receptors may also "knock-down" the supraspinal site of action for DPDPE. An antisense sequence (COR) derived from one of the highly conserved regions among the cloned opioid receptors produced a significant rightward shift in the antinociceptive response of opioid agonists without receptor type selectivity, such as etorphine, as well as of agonists with selectivity for μ (DAMGO), κ (U69,593 and bremazocine) and δ ([D-Ala²,Glu⁴]deltorphin and DPDPE).[30] The corresponding mismatch control ODN had no effect on the antinociceptive effect of any of the above named agonists. Supraspinal blockade of both [D-Ala²,Glu⁴]deltorphin and DPDPE by COR ODN suggests that the putative δ_1 and δ_2 receptors share a conserved nucleotide sequence that encodes residues 82-88 of the cloned δ receptor, but must differ at the start of the amino terminus. These findings imply that the differential effect of the δ antisense ODN on the antinociceptive response of [D-Ala²,Glu⁴]deltorphin and DPDPE is based on the ligands' differential interactions with two structurally related, but distinct receptors.

SEQUENCE-SPECIFIC "KNOCK-DOWN" OF DELTA OPIOID RECEPTORS IN NEUROBLASTOMA CELLS

The specificity by which the δ antisense ODN inhibit the synthesis of the cloned δ receptor has also been addressed in vitro. As mentioned above, NG 108-15 cells in cultured accumulate ODN in a time and concentration dependent manner. Cells which accumulate ODN are metabolically viable, as determined by the cells' ability to accumulate dextran subsequent to ODN treatment.[24] A polyclonal antibody specific for the cloned δ receptor was used to measure the level of the receptor, by immunocytochemical analysis, in cells that had been pretreated with the δ antisense ODN or the mismatch ODN.[24] It was found that the δ antisense ODN, but not the mismatch ODN, resulted in a significant reduction of the immunostaining intensity for the δ receptor in these cells. The reduction of the δ receptor density correlated well with the extent of ODN accumulation in individual cells, suggesting that the efficiency of the antisense ODN mediated "knock-down" of the δ receptor in these cells was critically dependent on the translocation of the ODN to its intracellular target. These data further substantiate the specificity of the antisense ODN mediated "knock-down" of its target by demonstrating a direct correlation between the localization of the ODN and the resultant reduction in the target protein.

CONCLUSIONS

We have briefly reviewed here our rationale and experimental approach in using antisense ODN to establish a molecular basis for δ receptor heterogeneity. Our findings are consistent with previous pharmacological evidence for the existence of at least two subtypes of δ opioid receptors; they also suggest that the currently established structure of the δ receptor may correspond to the pharmacologically defined $δ_2$ subtype. The structural identity of the so-called $δ_1$ receptor remains to be verified. In view of the pharmacological evidence with respect to the modulatory role of δ receptors in morphine analgesia and dependence, the therapeutic potential and low dependence liability of δ agonists, as well as other physiological roles of δ receptors including that in immune cells, progress in the molecular characterization of δ receptor heterogeneity continues to be an important objective in opioid receptor research.

REFERENCES

1. Helene C, Toulme JJ. Specific regulation of gene expression by antisense, sense and antigene nucleic acids. Biochim Biophys Acta 1990; 1049: 99-125.
2. Stein CA, Cheng YC. Antisense oligonucleotides as therapeutic agents—is the bullet really magical? Science 1993; 261:1004-1012.
3. Pilowsky PM, Suzuki S, Minson JB. Antisense oligonucleotides: a new tool in neuroscience. Clin Exp Pharmacol Physiol 1994; 21:935-944.
4. Wahlestedt C. Antisense oligonucleotide strategies in neuropharmacology. Trends Pharmacol Sci 1994; 15:42-46.
5. Heyman JS, Mulvaney SA, Mosberg HI et al. Opioid δ receptor involvement in supraspinal and spinal antinociception in mice. Brain Res 1987; 420:100-108.
6. Jiang Q, Mosberg HI, Porreca F. Antinociceptive effects of [D-Ala²]deltorphin II, a highly selective δ agonist in vivo. Life Sci Pharmacol Lett 1990; 47:PL-43-PL-47.
7. Jiang Q, Takemori AE, Sultana M et al. Differential antagonism of opioid delta antinociception by [D-Ala2, Leu5, Cys6]enkephaliln and 5'-naltrindole isothiocyanate (5'-NTII): evidence for delta receptor subtypes. J Pharmacol Exp Ther 1991; 257:1069-1075.

8. Portoghese PS, Sultana M, Takemori AE. Naltrindole 5'-isothiocyanate: a nonequilibrium, highly selective δ opioid receptor antagonist. J Med Chem 1990; 33:1547-1548.
9. Mattia A, Vanderah T, Mosberg HI et al. Lack of antinociceptive cross-tolerance between [D-Pen^2,D-Pen^5]enkephalin and [D-Ala^2]deltorphin II in mice: evidence for delta receptor subtypes. J Pharmacol Exp Ther 1991; 258:583-587.
10. Mattia A, Farmer SC, Takemori AE et al. Spinal opioid delta antinociception in the mouse: mediation by a single subtype of opioid delta receptor. J Pharmacol Exp Ther 1992; 260:518-525.
11. Vendarah T, Takemori, AE, Sultana M et al. Interaction of [D-Pen^2,D-Pen^5]enkephalin and [D-Ala^2,Glu^4]deltorphin with opioid delta receptor subtypes in vivo. Eur J Pharmacol 1994; 252:133-137.
12. Stewart PE, Hammond DL. Evidence for delta opioid receptor subtypes in rat spinal cord: studies with intrathecal naltriben, cyclic [D-Pen^2,D-Pen^5]enkephalin and [D-Ala^2,Glu^4]deltorphin. J Pharmacol Exp Ther 1993; 266:820-828.
13. Malmberg AB, Yaksh TL. Isobolographic and dose-response analyses of the interaction between intrathecal mu and delta agonists: effects of nalrindole and its benzofuran analog (NTB). J Pharmacol Exp Ther 1992; 263:264-275.
14. Tiseo PJ, Yaksh TL. Dose-dependent antagonism of spinal opioid receptor agonists by naloxone and naltrindole: additional evidence for delta-opioid receptor subtypes in the rat. Eur J Pharmacol 1993; 236:89-96.
15. Porreca F, Takemori AE, Portoghese PS et al. Modulation of mu-mediated antinociception by a subtyype of opioid δ receptor in the mouse. J Pharmacol Exp Ther 1992; 263:147-152.
16. Evans CJ, Keith DE, Morrison H et al. Cloning of the delta opioid receptor by functional expression. Science 1992; 258:1952-1955.
17. Kieffer BL, Befort K, Gaveriaux-Ruff C et al. The δ-opioid receptor: isolation of a cDNA by expression cloning and pharmacological characterization. Proc Natl Acad Sci 1992; 89:12048-12052.
18. Crooke ST. Therapeutic applications of oligonucleotides. Ann Rev Pharmacol Toxicol 1992; 32:329-376.
19. Reisine T, Bell GI. Molecular biology of opioid receptors. Trends Neurosci 1993; 16:506-510.
20. Black LE, Farrelly JG, Cavagnaro JA et al. Regulatory considerations for oligonucleotide drugs: updated recommendations for pharmacology and toxicology studies. Antisense Res Devel 1994; 4:299-301.
21. Ehrlich G, Patinkin D, Ginzberg D et al. Use of partially phosphorothioated "antisense" oligodeoxynucleotides for sequence-dependent modulation of hematopoiesis in culture. Antisense Res Devel 1994; 4:173-183.
22. Wahlestedt C, Pich EM, Koob GF et al. Modulation of anxiety and neuropeptide Y-Y1 receptors by antisense oligodoxynucleotides. Science 1993; 259:528-531.

23. Standifer KM, Chien CC, Wahlestedt C et al. Selective loss of delta opioid analgesia and binding by antisense oligodeoxynucleotides to a delta opioid receptor. Neuron 1994; 12:805-810.
24. Lai J, Crook T, Payne A et al. Antisense oligodeoxynucleotide-mediated reduction in the δ opioid receptors: direct correlation between antisense uptake and receptor "knock-down." Analgesia 1995; 1:535-538.
25. Loke SL, Stein CA, Zhang XH et al. Characterization of oligonucleotide transport into living cells. Proc Natl Acad Sci 1989; 86:3474-3478.
26. Yakubov LA, Deeva EA, Zarytova VF et al. Mechanism of oligonucleotide uptake by cells: involvement of specific receptors? Proc Natl Acad Sci 1989; 86:6454-6458.
27. Gao WY, Storm C, Egan W et al. Cellular pharmacology of phosphorothioate homo-oligodeoxynucleotides in human cells. Mol Pharmacol 1993; 43:45-50.
28. Beltinger C, Saragovi HU, Smith RM et al. Binding, uptake, and intracellular trafficking of phosphorothioate-modified oligodeoxynucleotides. J Clin Invest 1995; 95:1814-1823.
29. Sommer W, Bjelke B, Ganten D et al. Antisense oligonucleotide to c-fos induces ipsilateral rotational behaviour to d-amphetamine. NeuroReport 1993; 5:277-280.
30. Bilsky EJ, Berstein RN, Hruby VJ et al. Characterization of antinociception to opioid receptor selective agonists following antisense oligodeoxynucleotide-mediated "knock-down" of opioid receptors in vivo. J Pharmacol Exp Ther 1996; in press.
31. Lai J, Bilsky EJ, Rothman RB et al. Treatment with antisense oligodeoxynucleotide to the opioid δ receptor selectively inhibits $δ_2$-agonist antinociception. NeuroReport 1994; 5:1049-1052.
32. Bilsky EJ, Bernstein RN, Pasternak GW et al. Selective inhibition of [D-Ala^2,Glu^4]deltorphin antinociception by supraspinal, but not spinal, administration of an antisense oligodeoxynucleotide to an opioid delta receptor. Life Sci 1994; 55:PL-37-PL-43.

CHAPTER 4

FUNCTIONAL EFFECTS OF ANTISENSE OLIGODEOXYNUCLEOTIDES TO OPIOID RECEPTORS IN RATS

Jill U. Adams, Xiao-Hong Chen, J. Kim DeRiel, Jinling Yin, Martin W. Adler and Lee-Yuan Liu-Chen

Opioids such as morphine have effects on a number of physiological functions, including nociception, thermoregulation, respiration, gastrointestinal motility and immune function. With the characterization of three major opioid receptor types, μ, κ and δ,[1,2] and the development of selective agonists and antagonists for each, both cross-tolerance[3,4] and antagonism[5,6] studies have provided strong pharmacologic evidence for μ, κ and δ receptors playing distinct roles in opioid effects. A new method of manipulating opioid receptors, that is with antisense oligodeoxynucleotides (oligos), became possible when molecular cloning of μ, κ and δ receptors was achieved.[7]

Antisense oligos are thought to bind specifically, by complementary base-pairing, to targeted mRNA in cells. Once bound, they inhibit protein synthesis either by interfering with ribosomal translation or by increasing ribonuclease H degradation of the transcript.[8] By interrupting the synthesis of receptor proteins, antisense oligos should upset the homeostatic balance of receptor density and should reduce receptor function. Indeed, when administered in vivo, antisense oligos targeted to neurotransmitter receptors have been shown to affect physiological and behavioral functions; concomitantly, a reduction in receptor number has been observed.[9,10] As new receptors are being cloned at a rate which outpaces the development of pharmacologic agents with

Antisense Strategies for the Study of Receptor Mechanisms,
edited by Robert B. Raffa and Frank Porreca. © 1996 R.G. Landes Company.

which to characterize them, antisense oligos are potentially very useful tools to study the biological functions of new receptors and receptor subtypes. However, a number of issues have yet to be resolved concerning the fate of antisense oligos administered in vivo. Little is known regarding the pharmacokinetics of injecting oligos into brain tissue or into the cerebral ventricles including the distribution of the oligo from the injection site, the diffusion of the oligo across cell membranes, the degradation of the oligo and the duration of action. These uncertainties loom large in neurotransmitter systems which have not been well characterized. Using a well described system, such as the opioid system, many of these issues can be indirectly evaluated as the utility of in vivo oligo treatment is tested.

In this chapter, we summarize a series of studies from our laboratories in which the effects of i.c.v. administration of antisense oligos targeted to μ and κ opioid receptors were determined on receptor function and binding in rats. Opioid receptor function was evaluated by testing the well characterized antinociceptive and thermoregulatory effects of receptor-selective opioid agonists. Preliminary results with a μ antisense oligo on [^{3}H]-[D-Ala^2-NMePhe^4-Gly^5-ol]enkephalin (DAMGO) binding in rat brain membranes are also presented. In short, we found selective reductions in opioid receptor function and binding after in vivo antisense treatment. Pharmacologically, antisense oligo treatment affects opioid agonists in a similar manner to selective opioid antagonists. The results are discussed in the context of the strong background of opioid pharmacology in these areas.

ANTISENSE OLIGO TREATMENT

Our design of the antisense (AS) oligo, targeted for the N-terminal region of rat μ- and κ-opioid receptors, was based on observations that the N-terminal domain represents one of the most divergent regions among the opioid receptor types and that antisense oligos against sequences close to the initiation codon have been found to be effective in vivo.[10] In addition to 18-mer antisense sequences, sense (S) and missense (MS) oligos were also synthesized. Whereas antisense oligos-contain sequences complementary to the targeted mRNA, sense oligos contain the identical sequence as the targeted mRNA. Missense oligos contain the same base compositions as the antisense oligos but in a different sequence or differ from the antisense oligos only in a few mismatches. The sequences were as follows: μ-AS-oligo, 5-GCCGGTGCTGCTGTCCAT-3' (complementary to nucleotide number 1-18 of the coding region);[11]
μ-S-oligo, 5'-ATGGACAGCAGCACCGGC-3';
μ-MS-oligo, 5'-GCGGCTGGTCCTCTCGAT-3';
κ-AS-oligo, 5-AATCTGGATGGGGGACTC-3' (complementary to nucleotide number 4-21 of the coding sequence);[12]
κ-S-oligo, 5'-GAGTCCCCCATCCAGATT-3';

κ-MS-oligo, 5'-GACTGGAGGCACTTGATG-3 or
5'-ATTGTGCAGGTGCGAGTC-3'.
The nucleotide sequences targeted are unique to the μ and κ opioid receptors, respectively, with no homology to other opioid receptors or any other cloned receptors in the GeneBank database.

Male Sprague Dawley rats were surgically implanted with cannulae in the lateral cerebral ventricle according to standard procedures in our laboratory.[13] The oligo treatment protocol consisted of three i.c.v. injections of artificial cerebrospinal fluid (aCSF) or 20 μg oligodeoxynucleotide to unrestrained rats on days 1, 3 and 5. On day 6, animals were tested in antinociceptive or thermoregulatory assays or were sacrificed for binding studies. The treatment parameters used were based on the reported turnover rate of opioid receptor proteins being 3-5 days[14] and doses were similar to previously reported ones.[8] Using a phosphorothioate oligo, reported to be a metabolically more stable analog than the native phosphodiester-linked oligos,[8] we found that alternate day injections were sufficient to see changes in opioid receptor function and binding. Phosphorothioate oligos have been shown to have half-lives of more than 24 hr in serum or CSF.[15] I.c.v. administration of phosphorothioate oligos in rats did not result in any overt toxicity.

ANTINOCICEPTION STUDIES

Our laboratory has used the cold water tail-flick test in rats and has demonstrated its usefulness in studying antinociceptive effects of μ, κ and δ opioid receptor agonists in rats.[13,16] Specifically, agonists selective for κ and δ opioid receptors, when administered i.c.v., are inactive in the commonly used heat tail-flick tests in rats.[17,18] Not only are κ and δ agonists effective antinociceptive agents in the rat cold water tail-flick test, their actions are selectively antagonized by κ and δ receptor antagonists.[19] The cold water tail-flick test thus permits investigation of differentially mediated effects among agents acting on the three opioid receptor types.

Antisense oligo treatment (a total of 60 μg i.c.v. over 5 days in rats) reduced opioid receptor function as defined by a diminished antinociceptive effect after agonist administration as compared to aCSF control treatment. The receptor selectivity of i.c.v. antisense oligo treatment was demonstrated using agonists selective for the μ, κ and δ opioid receptors in the rat cold water tail-flick test. Figure 4.1 shows the diminished antinociceptive effects of [NMePhe³,D-Pro⁴]morphiceptin (PL017), a selective μ opioid agonist, after treatment with the antisense oligo targeted to the μ opioid receptor (μ-AS-oligo) compared to aCSF.[20] Meanwhile the effects of the κ selective agonist spiradoline and the δ selective agonist BW373U86 were unaffected by treatment with μ-AS-oligo treatment. In a parallel experiment, spiradoline-induced antinociception was of a smaller magnitude after treatment with the antisense

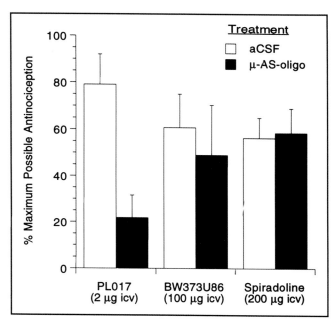

Fig. 4.1. Maximal possible antinociception after i.c.v. injection of PL017 (μ agonist), BW373U86 (δ agonist) or spiradoline (κ agonist) in rats treated with aCSF or μ-AS-oligo. Each bar represents the mean ± SE of 6-8 rats. Reprinted with permission from: Chen X-H, Adams JU, Geller EB et al. An antisense oligodeoxynucleotide to μ-opioid receptors inhibits μ-agonist-induced analgesia in rats. Eur J Pharmacol 1995; 275:105-108. © 1995 Elsevier Science B.V.

oligo targeted to the κ opioid receptor (κ-AS-oligo) compared to aCSF (Fig. 4.2).[21] No change in antinociceptive effect induced by PL017 or another δ selective agonist, [D-Pen^{2,5}]enkephalin (DPDPE) was observed after κ-AS-oligo treatment. Thus, μ-AS-oligo treatment selectively reduced the functionality of μ opioid receptors whereas κ-AS-oligo treatment selectively reduced κ opioid receptor function.

These findings demonstrated the successful use of antisense oligos in vivo to selectively inhibit opioid receptor mediated effects. Others have reported similar results using different oligo sequences, assays, injection sites and species. For example, Rossi et al[22] showed that a μ antisense oligo completely blocked the antinociceptive effects of morphine in rats. The μ antisense oligo used by Rossi et al[22] was directed against the 5'-untranslated region (68-87 bases 5' to the initiation codon), whereas our μ-AS-oligo was an 18-mer targeted at the first 18 nucleotides of the coding region of the μ opioid receptor. In addition, we administered oligos by i.c.v. injections and tested agonist effects in the cold water tail-flick test,[20] whereas they injected directly into the periaqueductal gray and used the radiant heat tail-flick assay.[22] In an-

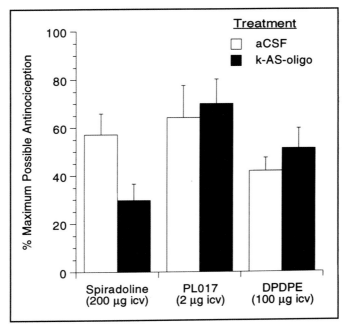

Fig. 4.2. Maximal possible antinociception after i.c.v. spiradoline (κ agonist), PL017 (μ agonist) and DPDPE (δ agonist) in rats treated with aCSF or κ-AS-oligo. Each bar represents the mean and SE of 4-9 rats. Reprinted with permission from: Adams JU, Chen X-H, DeRiel JK, et al. Intracerebroventricular treatment with an antisense oligodeoxynucleotide to κ-opioid receptors inhibited κ-agonist-induced-analgesia in rats. Brain Res 1994; 667:129-132. © 1994 Elsevier Science B.V.

other study, Chien et al[23] reported that i.t. administration of an antisense oligo against the mouse κ opioid receptor (complementary to nucleotides 577-597 of the coding region) selectively blocked the antinociceptive effects of the κ agonist U50,488H, a congener of spiradoline, but not those of μ or δ agonists, in mice.[16] Two groups have shown that antisense oligos against the mouse δ–opioid receptor (complementary to nucleotides 1-18 or [-2]-18 of the coding region) given i.t. attenuated the spinal antinociceptive effects of two δ agonists.[14,24] All the antisense oligos used were targeted to sequences unique to the receptor of interest, which accounts for their selectivity. Despite the specific differences in oligos, subjects and methods, the observations that antisense oligodeoxynucleotides selectively reduced opioid receptor function in each of these studies implies a general applicability of this approach for opioid receptors.

Other issues regarding opioid mediated antinociception have been addressed using antisense oligos. For instance, there has been some question regarding the site of action of κ receptor mediated antinociception. Our study with the κ-AS-oligo demonstrated that supraspinal sites are involved in the action of κ agonists in rats[21] and substantiated evidence reported in selective antagonist studies using hot-plate and cold-water tail-flick tests in rats.[19,25] Tseng and Collins[26] employed a δ antisense oligo to explore the mechanism of action of the endogenous opioid, β-endorphin. The antisense oligo was given i.t. and blocked antinociception induced by i.c.v. β-endorphin in mice, thus indicating spinal δ receptors were activated after supraspinal β-endorphin. These data supported similar findings from experiments with selective δ receptor antagonists administered i.t.[27,28] Bilsky et al[29] administered a δ antisense oligo i.c.v. and found a differential inhibition of supraspinal antinociception induced by the two δ agonists thought to act at different subtypes of the δ-opioid receptor. The same δ antisense oligo was found to attenuate antinociception induced by both δ agonists at the level of the spinal cord.[14] This evidence suggests different mechanisms for δ receptor mediated antinociception at spinal versus supraspinal loci. In addition, it suggests the existence of subtypes of the δ opioid receptor in mice even though only one molecular form has been cloned. Again, these data substantiate those of studies using selective δ receptor antagonists.[30,31]

Because treatment with antisense oligos is thought to reduce the number or density of functional receptors, a noncompetitive interaction with a receptor agonist, i.e. an insurmountable antagonism, might be predicted. Full agonist dose-effect curves are required to appropriately assess such an interaction. We used a cumulative dosing procedure so that an entire dose-effect curve for morphine or spiradoline was generated in each animal on the test day. Figure 4.3A shows the dose-effect curve for morphine after μ-AS-oligo treatment was shifted to the right compared to aCSF, μ-S-oligo or μ-MS-oligo treatment.[20] Higher doses of the μ-AS-oligo (i.e., 80 μg/i.c.v. injection or 240 μg total over 5 days) antagonized morphine-induced antinociception to an even greater extent (Fig. 4.3B). Similar results were obtained with spiradoline after κ-AS-oligo treatment and are displayed in Figure 4.4.[21]

To date, only one other in vivo antisense study has studied the effect on agonist dose-effect curves. Zhou et al[32] found a reduced maximum for quinpirole after treatment with antisense targeted for the D_2 dopamine receptor. This was consistent with the reduction in D_2 receptor number that these investigators found in binding experiments. In our experiments, the effect of i.c.v. antisense oligo treatment on s.c. morphine or spiradoline antinociception was characterized by a parallel shift to the right of the agonist dose-effect curve. This finding can be explained by the existence of spare receptors. That is, morphine may use only a small fraction of the μ-opioid receptor population to induce a maximal antinociceptive effect, therefore a large fraction of receptors would need to be inactivated to reduce the maximum

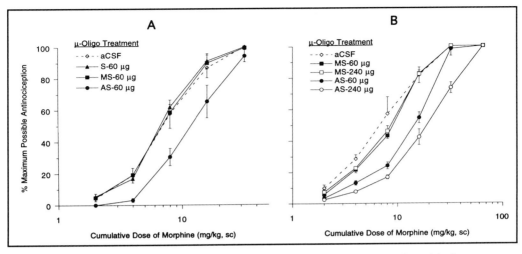

Fig. 4.3. Antinociceptive dose-effect curves for s.c. morphine (2-32 mg/kg, cumulatively) after treatment with aCSF or oligodeoxynucleotides (60 or 240 µg over 5 days) targeted to the µ-opioid receptor. Each curve represents mean data from 6-7 rats; vertical bars are SEs. Panel A reprinted with permission from: Chen X-H, Adams JU, Geller EB et al. An antisense oligodeoxynucleotide to µ-opioid receptors inhibits µ–agonist-induced analgesia in rats. Eur J Pharmacol 1995; 275:105-108. © 1995 Elsevier Science B.V.

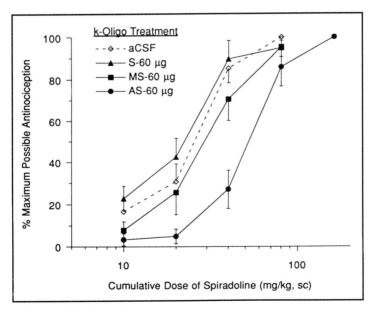

Fig. 4.4. Antinociceptive dose-effect curves for s.c. spiradoline (10-160 mg/kg, cumulatively) after treatment with aCSF or oligodeoxynucleotides targeted to the κ-opioid receptor. Each curve represents mean data from 7-9 rats; vertical bars are SEs. Reprinted with permission from: Adams JU, Chen X-H, DeRiel JK, et al. Intracerebroventricular treatment with an antisense oligodeoxynucleotide to κ-opioid receptors inhibited κ-agonist-induced-analgesia in rats. Brain Res 1994; 667:129-132. © 1994 Elsevier Science B.V.

possible effect. Experimentation with higher doses of μ-AS-oligo, 240 μg total over 5 days, produced a slightly larger shift of the morphine dose-effect curve compared to the 60 μg treatment, but still no diminution of maximum effect was observed (Fig. 4.3B). With the irreversible antagonist β-funaltrexamine (β-FNA), a reduction in maximum effect of morphine in the rat cold water tail-flick test is observed only after treatments that shift the dose-effect curve more than 5 times to the right (unpublished observations). Investigation with different antisense treatment protocols that might reduce receptor density sufficiently to clearly demonstrate a noncompetitive interaction with the opioid agonists is ongoing.

Taken together, these early results underscore the high degree of specificity that one can obtain with properly designed antisense oligodeoxynucleotides. The selective reduction in opioid receptor function induced by antisense oligo treatment in antinociceptive assays are very consistent with previous pharmacologic approaches to distinguishing among the opioid receptor types. Differential antagonism by the nonselective antagonists naloxone and naltrexone has been shown using pA_2 analysis.[4,33] Cross tolerance and lack thereof has often been used to confirm the selectivity of non-μ agonists such as U50,488H[4] and DPDPE.[3] The development of receptor-selective antagonists for μ-, κ- and δ-opioid receptors[5,6,34] allowed further corroboration of multiple opioid receptors mediating antinociception. Antagonists like β-FNA which selectively and irreversibly bind to opioid receptors have also been used to tease out receptor roles.[35] The mechanism of action of antisense oligos, thought to interfere with the synthesis of new receptors, differs significantly from competitive antagonism, irreversible antagonism and tolerance, all which appear to interfere with existing opioid receptors. The consistency of results among the different methods of manipulating opioid receptors serves to reinforce the large body of work characterizing differential receptor mechanisms mediating the antinociceptive effects of opioids.

THERMOREGULATION STUDIES

In addition to its role in nociception, the opioid system plays an important role in regulating body temperature (Tb).[36] Morphine, as well as more selective μ-opioid agonists such as PL017, increase core Tb in rats after i.c.v. administration or after direct injection into a brain area thought to be quite important in regulating Tb, the preoptic anterior hypothalamus.[37-39] This hyperthermia induced by μ agonists are blocked by μ-selective antagonists.[38] Intracerebroventricular injection of κ opioid receptor agonists, such as the endogenous opioid dynorphin A(1-17) and the selective ligand U50,488H, induces the opposite effect, i.e., a decrease in Tb, and this hypothermia is sensitive to κ-selective antagonists.[38-40] On the other hand, δ opioid receptors do not appear to be involved in the regulation of Tb.[38]

Antisense oligo treatment (a total of 60 µg i.c.v. over 5 days in rats) reduced opioid receptor function as defined by diminished agonist-induced effects on body temperature as compared to control. The receptor selectivity of i.c.v. antisense oligo treatment was demonstrated using selective µ- and κ-opioid agonists which induce opposite effects on Tb. The µ-selective agonist PL017 (1.0 µg i.c.v.) induced an increase in Tb of 2.5°C. Treatment with an antisense oligo targeted for the µ opioid receptor prevented the hyperthermic effect of PL017 compared to aCSF treatment (Fig. 4.5).[41] Neither µ-S-oligo nor µ-MS-oligo treatment altered the PL017-induced increase in Tb. The κ agonist dynorphin A(1-17) (10 µg i.c.v.) decreased Tb in rats by 1.7°C. Treatment with a κ-AS-oligo, but not with κ-S-oligo or κ-MS-oligo, attenuated the hypothermic effect of dynorphin A(1-17) (Fig. 4.6).[41] These results with antisense oligos are consistent with several reports using selective antagonists to differentiate between µ- and κ-opioid receptor mediated thermoregulatory effects.[36-40]

In addition to the possible existence of δ-opioid receptor subtypes discussed previously, subtypes of µ- and κ-opioid receptors have also been hypothesized and supporting evidence using selective irreversible antagonists has been reported.[42,43] However, as is the case with the δ-opioid receptor, different molecular forms of µ- and κ-opioid receptors

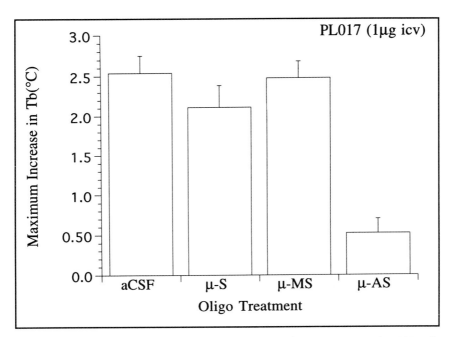

Fig. 4.5. Hyperthermic effect of PL017 (1.0 µg, i.c.v.) after treatment with aCSF, µ-S-, µ-MS- or µ-AS-oligo. Each bar represents the mean and SE in 6-8 rats.

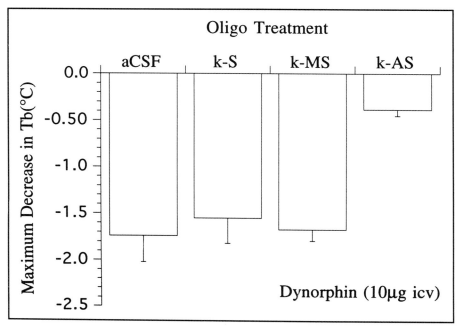

Fig. 4.6. Hypothermic effect of dynorphin A(1-17) (10.0 μg, i.c.v.) after treatment with aCSF, κ-S-, κ-MS- or κ-AS-oligo. Each bar represents the mean and SE in 5-8 rats.

have not yet been reported. If subtypes do exist at the pretranslational level, our data suggests that the subtypes of each receptor that have been cloned are apparently involved in the thermoregulatory effects of μ and κ agonists, as well as the antinociceptive effects. These results demonstrate again the successful use of antisense oligos in vivo to selectively inhibit opioid receptor mediated effects. To our knowledge, this was the first study to test the effects of antisense oligos against opioid receptors on a physiological response other than pain modulation. It further supports the general applicability of the antisense approach in opioid pharmacology.

RADIOLIGAND BINDING

To better understand the mechanism of the diminished opioid agonist effects after treatment with antisense oligos, opioid receptor binding studies were performed. After the same 5 day treatment protocol used in the antinociceptive and thermoregulatory experiments, animals were killed, brains were removed and P_2 membranes were prepared as described previously.[44] [^{3}H]DAMGO binding was conducted under conditions that converted all receptors to the high agonist affinity state.[45] Briefly, membranes were treated with 100 mM NaCl and 100 μM GDP. [^{3}H]DAMGO (1 nM) binding was performed in the presence of 5 mM Mg^{++}. (-)-Naloxone (1 μM) was used to define nonspecific binding.

Treatment with μ-AS-oligo (a total of 60 μg i.c.v. over 5 days in rats) reduced opioid receptor binding as defined by 1 nM [^{3}H]DAMGO binding by 20% compared to treatment with aCSF (Table 4.1). While the comparison between μ-AS- and μ-MS-oligo did not reach statistical significance, a similar reduction was observed. Whether this reduction in [^{3}H]DAMGO binding is due to a change in Kd and/or Bmax is currently being investigated.

The 20% decrease in opioid binding induced by the μ-AS-oligo treatment is reminiscent of the moderate reduction produced by an irreversible μ-opioid antagonist; a 30% reduction in [^{3}H]DAMGO binding was observed 24 hr after i.c.v. administration of 2 nmol β-FNA.[45] To further compare μ-AS-oligo to β-FNA, these same treatments both induced 2-fold parallel shifts of the morphine dose-effect curve in the rat cold water tail-flick test. The resemblance of antisense oligos to irreversible antagonists is not surprising because they both are thought to decrease the number or density of functional opioid receptors, albeit by different mechanisms. Antisense oligos as discussed are thought to inhibit synthesis of new receptors, whereas irreversible antagonists are thought to covalently bind to the agonist binding site of existing receptors.

Other studies have also found reduced receptor binding capacity of central nervous system tissue after in vivo antisense oligo treatment. In the mouse spinal cord, δ-opioid receptors were reduced by 25-30%.[14] Wahlestedt and colleagues[9,10] have studied nonopioid receptor systems and reported selective decreases in binding to NMDA-R1 and neuropeptide Y-Y1 receptors in rat brain after specific antisense oligo treatment.

CONCLUSIONS

The antisense oligos, and not in any case a sense or missense oligo, reduced opioid receptor function and binding. These findings are consistent with the putative mechanism of action of the antisense oligos,

Table 4.1. Effects of m-AS-oligo on [^{3}H]DAMGO binding to opioid receptors in rat brain

Treatment	n	dpm/nM [^{3}H]DAMGO/mg protein
aCSF	5	4370 ± 230
μ-MS-oligo	5	4304 ± 377
μ-AS-oligo	5	3523 ± 160 *

$P < 0.05$ compared to aCSF treatment group by Student's t test.

namely specific complementary base-pairing. The sense oligos, which have the same sequences as the target receptor mRNAs, should have no affinity for the targets. Likewise, the missense oligos, with the same base composition as the antisense oligos but in scrambled sequences or with only a few mismatches in the sequence, should not bind the targeted mRNA. Controls such as missense and sense oligos have previously been used with opioid and other neurotransmitter receptors to demonstrate specificity of in vivo antisense oligo effects on changes in behavioral endpoints, receptor densities and mRNA levels.[10,14,32,46]

While nonspecific interaction with cellular mRNA cannot be ruled out, there was no observable effect of any of the oligos on the animals. Body weights, baseline nociceptive latencies and baseline core body temperatures were not significantly different among rats treated with any of the oligos compared to those treated with aCSF. The finding that even the antisense oligo had no effect on baseline data is consistent with evidence from antagonist studies which generally report little to no effect of opioid receptor blockade on a wide variety of physiological measures. This supports the notion that opioids do not exert tonic control over nociception nor thermoregulation.

After cessation of antisense oligo treatment, the effects subside, presumably due to synthesis of new protein molecules. Standifer et al[14] showed that sensitivity to δ agonists returned to normal five days following the last i.t. injection of a δ-AS-oligo. In addition, 5-day but not 3-day treatment with the δ-AS-oligo significantly reduced DPDPE-induced antinociception.[14] The actions of antisense oligos are also dose-dependent. We demonstrated dose-dependent effects with our μ-AS-oligo in rats. Tseng et al[24] reported a similar dose-dependent attenuation of δ-induced antinociception by a δ-AS-oligo in mice.

Antisense oligo strategies offer several advantages over homologous recombination knockout strategies for in vivo study of gene products.[8] Because the effects of antisense oligos are transient and reversible after cessation of treatment, little or no compensatory mechanisms are involved. Antisense oligo treatment can be applied at any stage of development and to any species. The cost of antisense oligos for short-term in vivo treatment is much lower than breeding and raising animals and it permits more experimentation with various protocols. No sophisticated molecular cloning techniques are needed as long as the sequence of the molecule is known. Finally and importantly, use of antisense oligo strategies would be more relevant to possible therapeutic applications.

In summary, our studies demonstrate that antisense oligodeoxynucleotide treatment in vivo can reduce opioid receptor function and binding. These findings include the first report of antisense oligo effects on a nonnociceptive functional assay of opioids, namely thermoregulation. In the opioid system, at least three distinct receptors exist and antisense oligos can selectively inhibit the function of any one of them. Thus, the use of antisense oligos in vivo are comparable to that of selective receptor antagonists. Given the questions regarding the pharmacokinetics of oligodeoxynucleotide administration, our results together with those discussed above, when placed in the context of two decades of opioid receptor research, strongly support the viability of using antisense oligos in vivo. This work should be invaluable to promoting similar research in neurotransmitter systems that lack the pharmacologic tools of selective agonists and antagonists.

ACKNOWLEDGMENT

This work was supported by grants T32 DA07237, DA00376 and DA04745 from the National Institute on Drug Abuse.

REFERENCES

1. Lord JAH, Waterfield AA, Hughes J et al. Endogenous opioid peptides: multiple agonists and receptors. Nature 1977; 267:495-499.
2. Martin WR, Eades CG, Thompson JA et al. The effects of morphine- and nalorphine-like drugs in the nondependent and morphine-dependent chronic spinal dog. J Pharmacol Exp Ther 1976; 197:517-532.
3. Porreca F, Heyman JS, Mosberg HI et al. Role of mu and delta receptors in the supraspinal and spinal analgesic effect of [D-Pen^2,D-Pen^5]enkephalin in the mouse. J Pharmacol Exp Ther 1987; 241:393-400.
4. VonVoigtlander PF, Lewis RA. U-50,488H, a selective kappa opioid agonist: comparison to other reputed kappa agonists. Prog Neuro-Psychopharmacol Biol Psychiat 1982; 6:467-470.
5. Portoghese PS, Sultana M, Takemori AE. Naltrindole, a highly selective and potent nonpeptide ∂ opioid receptor agonist. Eur J Pharmacol 1988; 146:185-186.
6. Takemori AE, Ho BY, Naeseth JS et al. Nor-binaltorphimine, a highly selective kappa-opioid antagonist in analgesic and receptor binding assays. J Pharmacol Exp Ther 1988; 246:255-258.
7. Uhl GR, Childers S, Pasternak G. An opiate-receptor gene family reunion. Trends Neurosci 1994; 17:89-93.
8. Wahlestedt C. Antisense oligodeoxynucleotide strategies in neuropharmacology. Trends Pharmacol Sci 1994; 15:42-46.
9. Wahlestedt C, Golanov E, Yamamoto S et al. Antisense oligodeoxynucleotides to NMDA-R1 receptor channel protect cortical neurons from excitotoxicity and reduce focal ischaemic infarctions. Nature 1993; 363:260-263.
10. Wahlestedt C, Pich EM, Koob GF et al. Modulation of anxiety and neuropeptide Y-Y1 receptors by antisense oligodeoxynucleotides. Science 1993; 259:528-531.
11. Chen Y, Mestek A, Liu J et al. Molecular cloning and functional expression of a μ-opioid receptor from rat brain. Mol Pharmacol 1993; 44:8-12.
12. Li S, Zhu J, Chen G et al. Molecular cloning and expression of a rat κ opioid receptor. Biochem J 1993; 295:629-633.

13. Adams JU, Tallarida RJ, Geller EB et al. Isobolographic superadditivity between delta and mu opioid agonists in the rat depends on the ratio of compounds, the mu agonist and the analgesic assay used. J Pharmacol Exp Ther 1993; 266:1261-1267.
14. Standifer KM, Chien C-C, Wahlestedt C et al. Selective loss of ∂ opioid analgesia and binding by antisense oligodeoxynucleotide to a ∂ opioid receptor. Neuron 1994; 12:805-810.
15. Campbell JM, Bacon TA, Wickstrom E. Oligodeoxynucleotide phosphorothioate stability in subcellular extracts, culture media, sera and cerebrospinal fluid. J Biochem Biophys Methods 1990; 20:259-267.
16. Tiseo PJ, Geller EB, Adler MW. Antinociceptive action of intracerebroventricularly administered dynorphin and other opioid peptides in the rat. J Pharmacol Exp Ther 1988; 246:449-453.
17. Tyers MB. A classification of opiate receptors that mediate antinociception in animals. Br J Pharmacol 1980; 69:503-512.
18. Heyman JS, Vaught JL, Raffa RB et al. Can supraspinal δ-opioid receptors mediate antinociception? Trends Pharmacol Sci 1988; 9:134-138.
19. Adams JU, Geller EB, Adler MW. Receptor selectivity of i.c.v. morphine in the rat cold water tail-flick test. Drug Alcoh Dep 1994; 35:197-202.
20. Chen X-H, Adams JU, Geller EB et al. An antisense oligodeoxynucleotide to μ-opioid receptors inhibits μ-agonist-induced analgesia in rats. Eur J Pharmacol 1995; 275:105-108.
21. Adams JU, Chen X-H, DeRiel JK et al. Intracerebroventricular treatment with an antisense oligodeoxynucleotide to κ-opioid receptors inhibited κ-agonist-induced-analgesia in rats. Brain Res 1994; 667:129-132.
22. Rossi G, Pan Y-X, Cheng J et al. Blockade of morphine analgesia by an antisense oligodeoxynucleotide against the μ receptor. Life Sci 1994; 21:PL375-PL379.
23. Chien C-C, Brown G, Pan Y-X et al. Blockade of U50,488H analgesia by antisense oligodeoxynucleotides to a κ-opioid receptor. Eur J Pharmacol 1994; 253:R7-R8.
24. Tseng LF, Collins KA, Kampine JP. Antisense oligodeoxynucleotide to a ∂-opioid receptor selectively blocks the spinal antinociception induced by ∂-, but not μ- or κ-opioid receptor agonist in the mouse. Eur J Pharmacol 1994; 258:R1-R3.
25. Jones DNC, Holtzman SG. Long term κ-opioid receptor blockade following nor-binaltorphimine. Eur J Pharmacol 1992; 215:345-348.
26. Tseng LF, Collins KA. Antisense oligodeoxynucleotide to a ∂-opioid receptor given intrathecal blocks i.c.v. administered β-endorphin-induced antinociception in the mouse. Life Sci 1994; 55:PL127-PL131.
27. Suh HH, Tseng LF. Delta but not mu-opioid receptors in the spinal cord are involved in antinociception induced by β-endorphin given intracerebroventricularly in mice J Pharmacol Exp Ther 1990; 253:981-986.
28. Tseng LF, Collins KA, Portoghese PS. Spinal δ_2 but not δ_1 opioid receptors are involved in intracerebroventricular β-endorphin-induced antinociception in the mouse. Life Sci 1993; 52:PL211-PL215.

29. Bilsky EJ, Bernstein RN, Pasternak GW et al. Selective inhibition of [D-Ala², Glu⁴]deltorphin antinociception by supraspinal, but not spinal, administration of an antisense oligodeoxynucleotide to an opioid delta receptor. Life Sci 1994; 55:PL37-PL43.

30. Sofuoglu M, Portoghese PS, Takemori AE. Differential antagonism of delta opioid agonists by naltrindole and its benzofuran analog (NTB) in mice: evidence for delta opioid receptor subtypes. J Pharmacol Exp Ther 1991; 257:676-680.

31. Mattia A, Farmer SC, Takemori AE et al. Spinal opioid delta antinociception in the mouse: mediation by a 5'-NTII-sensitive delta receptor subtype. J Pharmacol Exp Ther 1992; 260:518-525.

32. Zhou L-W, Zhang S-P, Qin Z-H et al. In vivo administration of an oligodeoxynucleotide antisense to the D_2 dopamine receptor messenger RNA inhibits D_2 dopamine receptor-mediated behavior and the expression of D_2 dopamine receptors in mouse striatum. J Pharmacol Exp Ther 1994; 268:1015-1023.

33. Tung AS, Yaksh TL. In vivo evidence for multiple opiate receptors mediating analgesia in the rat spinal cord. Brain Res 1982; 247:75-83.

34. Kramer TH, Shook JE, Kazmierski W et al. Novel peptidic mu opioid antagonists: pharmacologic characterization in vitro and in vivo. J Pharmacol Exp Ther 1989; 249:544-551.

35. Takemori AE, Larson DL, Portoghese PS. The irreversible narcotic antagonistic and reversible agonistic properties of the fumarate methyl ester derivative of naltrexone. Eur J Pharmacol 1981; 70:445-451.

36. Adler MW, Geller EB. Physiological functions of opioids: temperature regulation. In: Herz A, Akil H, Simon EJ, eds. Handbook of Experimental Pharmacology. Vol 104/II, Opioids II. Berlin: Springer-Verlag, 1992:205-238.

37. Geller EB, Rowan CH, Adler MW. Body temperature effects of opioids in rats: intracerebroventricular administration. Pharmacol Biochem Behav 1986; 24:1761-1765.

38. Handler CM, Geller EB, Adler MW. Effect of μ-, κ-, and ∂-selective opioid agonists on thermoregulation in the rat. Pharmacol Biochem Behav 1992; 43:1209-1216.

39. Spencer RL, Hruby VJ, Burks TF. Body temperature response profiles for selective mu, delta, and kappa opioid agonists in restrained and unrestrained rats. J Pharmacol Exp Ther 1988; 246:92-101.

40. Cavicchini E, Candeletti S, Ferri S. Effects of dynorphins on body temperature of rats. Pharmacol Res Commun 1988; 20:603-604.

41. Chen X-H, Geller EB, DeRiel JK et al. Antisense oligodeoxynucleotides against μ- or κ-opioid receptors block agonist-induced body temperature changes in rats. Brain Res; in press.

42. Pasternak GW, Wood PJ. Minireview: multiple mu opiate receptors. Life Sci 1986; 38:1889-1898.

43. Horan PJ, DeCosta BR, Rice KR et al., Differential antagonism of U69,593- and bremazocine-induced antinociception by (-)-UPHIT:

evidence of kappa opioid receptor multiplicity in mice. J Pharmacol Exp Ther 1991: 257:1154-1161.
44. Liu-Chen L-Y, Phillips CA. Covalent labeling of μ opioid binding site by [^{3}H]β-funaltrexamine. Mol Pharmacol 1987; 32:321-329.
45. Liu-Chen L-Y, Yang H-H, Li S et al. Effect of intracerebroventricular β-funaltrexamine on μ opioid receptors in the rat brain: consideration of binding condition. J Pharmacol Exp Ther 1995; 273:1047-1056.
46. Zhang M, Creese I. Antisense oligodeoxynucleotide reduces brain dopamine D_2 receptors: behavioral correlates. Neurosci Lett 1993; 161:223-226.

CHAPTER 5

IN VIVO ANTISENSE STRATEGY FOR THE STUDY OF SECOND-MESSENGERS: APPLICATION TO G-PROTEINS

Robert B. Raffa

INTRODUCTION

The use of an in vivo antisense strategy in which the receptor serves as the target has been applied with great success (e.g., refs. 1-5 and other chapters of this book). It was our purpose to extend the logic of this approach to the next level in the receptor-effector transduction pathway—i.e., to the level of second messenger coupling. Specifically, we were interested in targeting G-proteins.

G-proteins are involved in a large number of receptor-mediated events and multiple GPCR (G-protein-coupled receptor) families are known in prokaryotes and eukaryotes.[6,7] G-proteins serve as intermediaries in the transduction of ligand-binding signals that lead to the activation or inhibition of second-messenger pathways and ion channels (Fig. 5.1). The magnitude of the signal transmitted by the G-proteins is ligand-dependent, i.e., binding of ligand to the receptor results in an increase or decrease in second-messenger or ion channel activity. In order to accomplish this ligand-dependent function, G-proteins cycle through an activated phase initiated by the ligand binding to the receptor, followed by an association with the receptor and the substitution of GTP for GDP, and subsequent return upon the hydrolysis of GTP by GTPase contained within the G-protein.[8] G-proteins are

Antisense Strategies for the Study of Receptor Mechanisms, edited by Robert B. Raffa and Frank Porreca. © 1996 R.G. Landes Company.

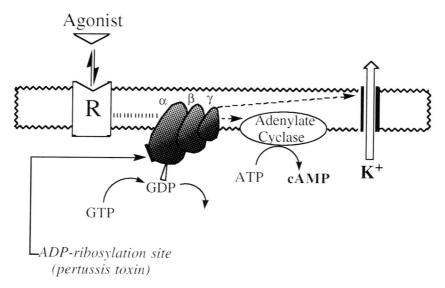

Fig. 5.1. Schematic representation of some aspects of G-protein function. In the latent state (nonbound receptor), the α and β/γ subunits are in an associated configuration and bound to GDP. Upon agonist binding to the receptor, GDP is displaced by GTP and the α subunit dissociates to stimulate (G_s) or inhibit (G_i) second messenger pathways (e.g., adenylate cyclase) or ion channels. The β/γ subunit assembly also affects transduction systems, possibly inhibiting heterologous receptor activity. The effect of antagonist-receptor binding on G-protein function is less well known. Pertussis toxin facilitates ADP-ribosylation of the α subunit of G_i proteins.

heterotrimeric, comprised of components designated α, β and γ. The resting-state consists of a complex of α, β and γ units (and a GDP molecule attached to the α unit) that is dissociated from the receptor. Activation occurs upon agonist binding to the receptor, GDP-α/βγ association with the receptor (at a site presumably different from the agonist binding site), GDP-α uncoupling from βγ, and the initiation by GTP-α of the next step in the signal transduction pathway. Multiple pathways can be coupled to G-proteins and evidence suggests that intracellular "cross-talk" can occur among the pathways. For example, heterogeneity of G-protein subunits can allow different receptors to exert opposite effects on a target enzyme. Hence, G-protein activation can be an important aspect of cellular control mechanisms.

G-proteins have been subdivided into categories, originally on the basis of their ability to stimulate or inhibit (G_s and G_i, respectively) adenylyl cyclase, and now on molecular biological bases. Multiple subunits of G-proteins exist,[9] allowing for a diversity of biological responses. This multiplicity, and the diversity of responses, offers targets ripe for fruitful investigation. We chose to examine the relationship between G-protein subunits in the mediation of antinociception (analgesia) and some side-effects produced by opioids (such as morphine) and nonopioids (α_2-adrenoceptor agonists).

G-PROTEIN ANTISENSE

The major focus of the work that is discussed in this chapter is the α subunit component of G-proteins and some biological effects mediated by it. There exist multiple subunits of mammalian G-protein α subunits, consisting of at least $G_s\alpha$, $G_i\alpha$, $G_q\alpha$, and $G_{12/13}\alpha$ groups (Fig. 5.2). We elected to examine $G_s\alpha$, $G_i1\alpha$, $G_i2\alpha$ and $G_i3\alpha$ on the basis of the availability of specific antisense sequences to each of these subunits and the extensive demonstration and characterization of their selectivity and effectiveness in vitro.[10,11] The individual nucleotide sequences and the amino acid correspondence of each of the antisense oligodeoxynucleotides (oligos)[10-14] are given in Figure 5.3. In each case, the antisense was designed from a 10 amino acid sequence of the G-protein subunit and, thus, each oligo was a 33-mer. Previous work by other investigators has shown that oligos of this length are capable of producing effects in vivo.[1] To enhance the biological stability (half-life) of the oligos, phosphorothioate analogs (Fig. 5.4) were utilized in

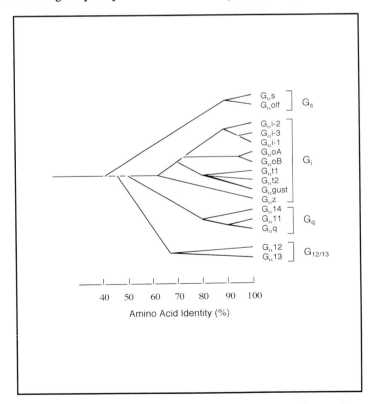

Fig. 5.2. α Subunits of mammalian G-protein. Courtesy of Santa Cruz Biotechnology (Santa Cruz, CA).

	a.a.		Oligonucleotide Sequences
(A)			
$G_i1\alpha$	134–124	s	3´-CGACAGGAAGGTGTCAGAGAAATACTGCGGCCG-5´
		as	5´-GCTGTCCTTCCACAGTCTCTTTATGACGCCGGC-3´
$G_i2\alpha$	135–125	s	3´-TACCAGTCGGGTCTCGGAGGCCTACTGCGGGCT-5´
		as	5´-ATGGTCAGCCCAGAGCCTCCGGATGACGCCCGA-3´
$G_i3\alpha$	134–124	s	3´-CGGTAGAGCGGTATTTGCAAATTAGTGCGGACG-5´
		as	5´-GCCATCTCGCCATAAACGTTTAATCACGCCTGC-3´
$G_s\alpha$	157–147	s	3´-GAGTAGGAGGGTGTCTCGGAACCGTACGAGTAT-5´
		as	5´-CTCATCCTCCCACAGAGCCTTGGCATGCTCATA-3´

From refs. **10** and **11**.

(B)			
$G_i1\alpha$	134–124	R	3´-CGA CAG GAA GGT GTC AGA GAA ATA CTG CGG CG-5´
	135–125	M	3´-<u>TGG</u> CAG GAA GGT GT<u>T</u> <u>GAA</u> <u>GGA</u> <u>C</u>TA CTG C<u>AG</u> <u>GTG</u>-5´
$G_i2\alpha$	135–125	R	3´-TAC CAG TCG GGT CTC GGA GGC CTA CTG CGG GCT-5´
	135–125	M	3´-<u>C</u>AC CAG TCG GGT CTC GGA GGC CTA CTG <u>TGG</u> <u>C</u>CT-5´
$G_i3\alpha$	134–124	R	3´-CGG TAG AGC GGT ATT TGC AAA TTA GTG CGG ACG-5´
		M	?
$G_s\alpha$	157–147	R	3´-GAG TAG GAG GGT GTC TCG GAA CCG TAC GAG TAT-5´
	140–130	M	3´-GAG TAG GAG GGT GTC TCG GAA CCG TAC GAG TAT-5´

From refs. **12–14**.

Fig. 5.3. *Nucleotide sequences of the G-protein antisense oligos utilized in the studies discussed in this chapter. (A) The antisense (as) was designed to interact with the sense (s) sequences of the amino acids (a.a.) indicated for each of the subunits. (B) Comparison of rat (R) and mouse (M) nucleotide sequences. Differences are underlined.*

the early studies. However, dosing of these was limited by their toxicity (including barrel-rolling behavior and pre-convulsive activity). We subsequently found that unmodified phosphodiester oligos, although slightly less potent than their phosphorothioate equivalents, could be given at higher doses without producing behavioral effects.

EXPERIMENTAL DESIGNS

All of the experiments described in this chapter relate to work on supraspinally-mediated antinociception, i.e., produced by direct intracerebroventricular (i.c.v.) injection of agonist, and measured using a standard mouse tail-flick test.[15] Features of the experiments that are common to all are described here. Differences are described under the specific experimental protocols.

Male, 18-24 g, virus-free albino Crl:CD-1®(ICR) mice (Charles

In Vivo: Application to G-Proteins

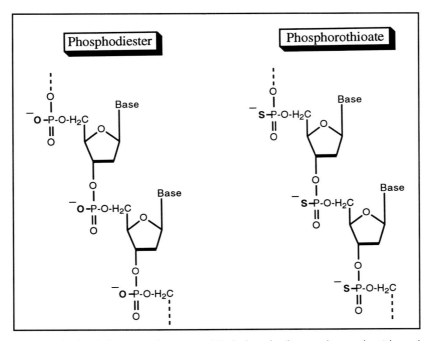

Fig. 5.4. The basic features of an unmodified phosphodiester oligonucleotide and phosphorothioate analog.

River Laboratories) were used in all of the experiments. They were housed 5-10 per plastic container in a climate-controlled, virus-free environment and acclimated for at least five days prior to testing. Food and water were available ad libitum up to the time of the testing. All of the mice were treated in accordance with policies and recommendations of the National Institutes of Health (NIH) and Johnson & Johnson Guidelines for the care and use of laboratory animals. The antisense oligos were purchased from Midland Certified Reagent Co. (Midland, TX), Cruachem Inc. (Sterling, VA) or were synthesized by J. R. Chambers and A. Bittner (The R. W. Johnson Pharmaceutical Research Institute, La Jolla, CA). The other compounds were purchased from Peninsula Laboratories (Belmont, CA), Research Biochemicals, Inc. (Natick, MA) and Sigma Chemical Co. (St. Louis, MO). All of the compounds were administered intracerebroventricularly into the lateral cerebral ventricle (total volume of 5μL) according to the method of Hylden and Wilcox.[16] Antinociception was measured as the increase in reaction time (latency) in the 55°C water tail-flick test[15] and was expressed as the percent of the maximum possible effect (% MPE) according to the equation: %MPE = 100 x (TL − CL)/(15 s − CL), where CL is the control latency (pre-drug) and TL is the test latency (post-drug) of each of the animals and 15 s is the maximal time of exposure. Antinociceptive effect is reported as the group mean (± s.e.m.) of % MPE values at the time of peak antinociceptive effect for each agonist. Typically, N = 10 mice per group. In all cases, statistical significance was tested with one-way ANOVA and, if P < 0.05, post-hoc test (Bonferroni multiple comparisons).

G-PROTEINS AND ANTINOCICEPTION

Pertussis toxin (PTX) and cholera toxin (CTX) catalyze ADP-ribosylation of the α subunit of G_i and G_s G-proteins, respectively, causing persistent activation (G_s) or inhibition (G_i). PTX pretreatment attenuates i.c.v. morphine-induced antinociception in mice and CTX enhances it.[17] Likewise, treatment with G_s antisera enhances, and G_i antisera attenuates, i.c.v. morphine-induced antinociception.[18,19] We have extended these observations by showing that the magnitude of the PTX effect is in inverse order of the efficacy of the agonist. That is, the effect is most pronounced for morphine, less for [D-Ala^2,NMePhe^4,Gly-ol^5]enkephalin (DAMGO), and least for sufentanil (manuscript in preparation), compounds of low, medium, and high efficacy, respectively.[20]

IN VIVO G-PROTEIN ANTISENSE

OPIOID ANTINOCICEPTION

Our first study that utilized an in vivo antisense strategy involved the examination of the functional interaction of supraspinal μ opioid receptors with the G protein subunits $G_i1α$, $G_i2α$, $G_i3α$ and $G_sα$.[20] Mice were injected i.c.v. with 33-base phosphorothioate oligodeoxynucleotides (12.5 μg) or vehicle in equal volume (sterile water, 5 μL) and the antinociceptive response to i.c.v. morphine was determined 18-24 h later using the tail-flick test. The results are shown in Figure 5.5. There was no difference (P > 0.05) in the antinociceptive effect in animals pretreated with $G_i1α$, $G_i3α$ or $G_sα$ antisense compared to vehicle-treated controls. However, pretreatment with $G_i2α$ antisense produced a significant reduction (> 70%; P < 0.05) of morphine-induced antinociception. These findings suggested: (1) that application of an in vivo antisense approach could detect a G-protein-dependent (PTX-sensitive) response; and (2) that a particular G-protein subunit (i.e., $G_i2α$) was involved to a greater degree than other subunits in the antinociceptive response. The latter result is in agreement with the report that i.c.v. administration of anti-$G_i2α$, but not of anti-$G_i1α$, immune sera reduces the supraspinal antinociceptive effect of morphine in the tail-flick test in mice.[19] We extended these findings by examining the effect of the same antisense treatment on opioid agonists of differing efficacies.[20] As was the case for morphine (low efficacy), pretreatment with $G_i1α$, $G_i3α$ or $G_sα$ antisense had no effect on DAMGO (medium efficacy) or sufentanil (high efficacy) antinociception. However, DAMGO-induced antinociception was reduced 43.6% by pretreatment with $G_i2α$ antisense and sufentanil-induced

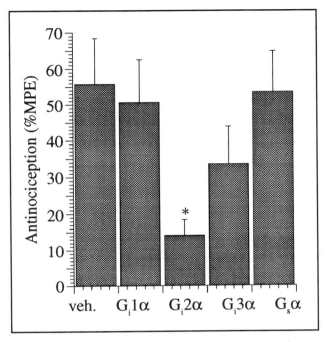

Fig. 5.5. Antinociception in the tail-flick test (expressed as % MPE) following the i.c.v. injection of morphine to mice. At 18-24 h prior to testing, the mice were injected i.c.v. with vehicle (veh.) or with antisense (12.5 µg) to $G_i1\alpha$, $G_i2\alpha$, $G_i3\alpha$ or $G_s\alpha$. N = 10 mice per group. The asterisk denotes significant difference (P < 0.05) from the vehicle-treated group. Reprinted with permission from: Raffa RB, Martinez RP, Connelly CD. Eur J Pharmacol 1994: 258:R5-7.

antinociception was not significantly affected. Hence, the degree of attenuation by oligo of the three agonists was in the order morphine > DAMGO > sufentanil, the inverse order of their efficacy, and consistent with the results that we obtained using PTX. Together, these data suggest that the supraspinal antinociceptive effect of µ opioid agonists is related to an interaction with the $G_i2\alpha$ subunit. The results further suggest that the efficacy of µ opioid agonists might be determined by the degree of utilization of the $G_i2\alpha$ pathway, because the order of antisense effect on morphine > DAMGO > sufentanil is the same as the order of increasing intrinsic efficacy.[21]

Perhaps the most intriguing aspect of the results of this first study was the possibility that opioid-induced antinociception might be mediated through only one G-protein α subtype, but that opioid side-effects or abuse liability might be mediated through a different G-protein α subtype; or, another possibility, that the antinociception induced by nonopioids might be mediated through a different G-protein α subunit than that utilized by opioid analgesics. In either case, it might be possible to selectively target G-protein subunits for particular effects.

TIME-COURSE OF ANTISENSE EFFECT

As in the study described above, we have generally found that a single injection of antisense oligo is sufficient to produce a significant effect. In this respect, our studies differ from the earlier protocols, in which multiple injections of oligos were used, typically for 3 or more days—based on an estimated turnover rate of receptors of 3-5 d. The surprising finding that a single antisense injection was sufficient was pursued in a subsequent study, which more thoroughly examined the time-course of the effect of $G_i2\alpha$ antisense oligo on μ-opioid-induced antinociception.[22]

The general protocol was similar to that described above. Mice were injected i.c.v. with 33-mer oligodeoxyribonucleotide (6.0 nmol) directed against $G_i1\alpha$ or $G_i2\alpha$ subunits, or with vehicle, and the antinociceptive response to i.c.v. μ-opioid agonist (morphine, DAMGO or sufentanil) was determined 0.75 h to 8 d later using the tail-flick test. Similar to the first study, the results showed a significant and selective attenuation of morphine-induced antinociception by $G_i2\alpha$ oligo, but not by $G_i1\alpha$ oligo or by vehicle. The attenuation was apparent by 1.5 h and persisted up to about 48 h after the injection of oligo (Fig. 5.6). Also similar to the first study, there was a significant and selective oligo-induced attenuation of DAMGO-induced antinociception by $G_i2\alpha$, but not $G_i1\alpha$ or vehicle. As was the case for morphine, the attenuation was apparent at 1.5 h and persisted up to about 24 h after the oligo injection. The onset of oligo attenuation of antinociception was about the same for morphine and DAMGO, but the duration of effect was slightly longer for morphine (48 h vs. 24 h). The magnitude of the attenuation of DAMGO-induced antinociception was less than the attenuation of morphine-induced antinociception throughout the duration of effect. Sufentanil-induced antinociception was minimally affected by oligo pretreatment. In each case, there was an apparent slight vehicle effect early in the time-course. The relative effects were evident when each response was normalized to the vehicle effect (Fig. 5.7). Confirming and extending the earlier study,[20] the attenuation by oligo was more pronounced for morphine (the lower efficacy agonist) than it was for DAMGO (the medium efficacy agonist) or sufentanil (the high efficacy agonist) throughout the duration of the effect.

In addition to the specific conclusion that opioid antinociceptive efficacy is inversely related to sensitivity to G-protein oligo antisense, the rapidity of the onset and the long duration of a single oligo administration are intriguing. The rapidity of the onset of effect suggests a rapid turnover of the α subunits of G-proteins, which, given the nature of their role in signal transduction, is not (in retrospect) surprising. An explanation of the long duration of the effect awaits

In Vivo: Application to G-Proteins 61

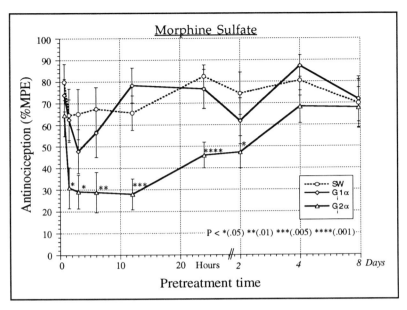

Fig. 5.6. Antinociception in the tail-flick test (expressed as % MPE) following the i.c.v. injection of morphine to mice following prior i.c.v. injection of vehicle or $G_i1\alpha$ or $G_i2\alpha$ antisense (6.0 nmol). N = 10-25 mice per group. The asterisks denote a significant difference from the vehicle-treated group. Reprinted with permission from: Stone Jr DJ, Wild KD, Raffa RB. Analgesia 1995; 1:770-773.

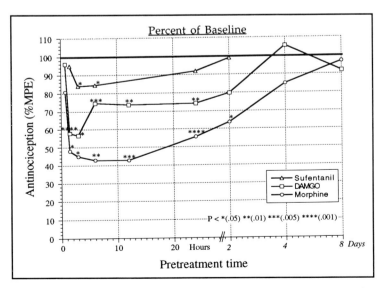

Fig. 5.7. Data from Fig. 5.7 normalized to the vehicle-treated group. The asterisks denote significant difference from baseline. Reprinted with permission from: Stone Jr DJ, Wild KD, Raffa RB. Analgesia 1995; 1:770-773.

further elucidation, including determination of the half-life of antisense within the relevant cells or cell compartment(s). Another aspect of this question is the effect of antisense on receptor number and affinity for ligand. To address this issue, we conducted preliminary radioligand-binding and autoradiographic examinations of the brains of mice treated with G-protein antisense using [^{3}H]DAMGO, an agonist that should be sensitive to decreased numbers of G-protein coupled receptors (high-affinity state), but should not detect uncoupled receptors. The preliminary findings are that there is no change in affinity for ligand (K_d), but that there might be an increase in receptor number (B_{max}). The latter was supported by autoradiographic findings using [^{3}H]-DAMGO (Fig. 5.8). These are preliminary findings and should be interpreted with caution. However, they suggest that G-protein-coupled receptors are slightly up-regulated during functional interference with one α subtype, and that other α subtypes substitute, thus maintaining constant, or slightly higher, the number of receptors that are coupled to G-proteins in the high-affinity state. In this model, the different α subtypes must be in equilibrium with the receptor, each having different affinities for it. In the absence of one, the others are free to interact. However, the substituted subunits are less able to mediate the same effect. Hence, the unusual situation of attenuated response and increased number of G-protein-coupled receptors. Again, however, these results await confirmation. Indeed, in a follow-up study, there was no difference in the radioligand binding parameters (K_d or B_{max}) of whole-brain tissue for [^{3}H]DAMGO, but this is still interesting, since there was no decrease detected. Future work will concentrate on the specific brain regions suggested by the autoradiographic results.

OPIOID SIDE-EFFECTS

Once we knew that morphine-induced antinociception was mediated via the $G_i2α$ subunit, it was immediately of interest to determine if morphine's side-effects, such as constipation, and its dependence liability are mediated by the same subunits. In this series of experiments, antisense (6 nmol) or vehicle (5 μL) was injected 3 h prior to i.c.v. morphine (10 μg) and antinociception was measured as before, over 4 h. The same mice were then injected with naloxone hydrochloride (10 mg/kg, i.p.). Acute dependence was quantified as the percent of mice per group that jumped[23,24] at least once over a 10 min period and, for responders, the number of jumps. Constipation was measured as the increase in colonic glass-bead retention time.[25] The results are shown in Figure 5.9. As found in our previous study,[20] the $G_i2α$ antisense oligo produced a significant reduction in morphine-induced antinociception (A). However, there was no significant alteration of acute dependence-liability (as measured by naloxone-precipitated jumping) or of constipation (B).[26] The finding of a lack of effect of antisense to $G_i2α$ on constipation or acute dependence—in the same animals in which it attenuated antinociception—implies that either the $G_i2\alpha$ subunit is not a major component of the transduction process(es) leading to constipation and acute dependence or that these effects are less sensitive to $G_i2\alpha$ antisense. That is, morphine could have different intrinsic activities at different endpoints. The implication of this finding is that opioid-induced analgesia and side-effects are separable phenomena at the level of G-protein α subunits and that novel analgesics could be designed by targeting particular subunits.

NONOPIOID ANTINOCICEPTION

Given the above findings that μ-opioid-induced supraspinal antinociception appeared to be preferentially mediated via one particular G-protein subunit ($G_i2\alpha$), it was of subsequent interest to determine if supraspinal antinociception induced by nonopioids utilized the same subunit. To investigate this question, we examined the effect of G-protein antisense on i.c.v. antinociception produced by α_2-adrenoceptor agonists (clonidine, guanfacine, and BH-T 920). α_2-Adrenoceptor agonist-induced supraspinal antinociception is known to be mediated through G-proteins sensitive to ADP-ribosylation by pertussis toxin.[27] For these experiments,[28] mice were injected with oligos (6 nmol) or vehicle and were tested (tail-flick) with an agonist administered i.c.v. 18-24 h later. $G_i2\alpha$ antisense produced differential effects on the three agonists. Unlike the studies with opioids, the $G_i3\alpha$ antisense treatment attenuated BH-T 920 and clonidine-induced antinociception (Fig. 5.10). Similar to the results with opioids, $G_i1\alpha$ and $G_s\alpha$ antisense oligo treatment had no significant effect. Interestingly, the attenuation of antinociception by the G_i antisense oligos was qualitatively in the order $G_i3\alpha > G_i2\alpha \approx G_i1\alpha$, the same order as reported by Kurose et al[29] in studies in vitro of $\alpha2$-C10 and α_2-C4 adrenoceptors reconstituted in phospholipid vesicles. Taken together with the prior findings, the results of this series of experiments suggest that multiple G-protein subunits might mediate supraspinal antinociception in mice and, further, that the subunit and receptor type might be preferentially coupled (viz., $G_i3\alpha$ with α_2-adrenergic and $G_i2\alpha$ with μ-opioid receptors). Of particular interest to us is that novel therapeutic agents might result from targeting "nonopioid" G-protein subunits.

CAVEATS

The same general concerns that apply to the use of receptor antisense in vivo apply to the studies of antisense directed at G-proteins. There is the usual concern about the in vivo specificity of the effect of antisense pretreatment, regardless of the in vitro specificity. Is the effect due to unrecognized toxicity or interaction of oligo with a nontargeted site? Unambiguous controls are not always readily apparent. Hence, the question of what constitutes proper controls for in vivo antisense experiments is the subject of much discussion and debate. The use of

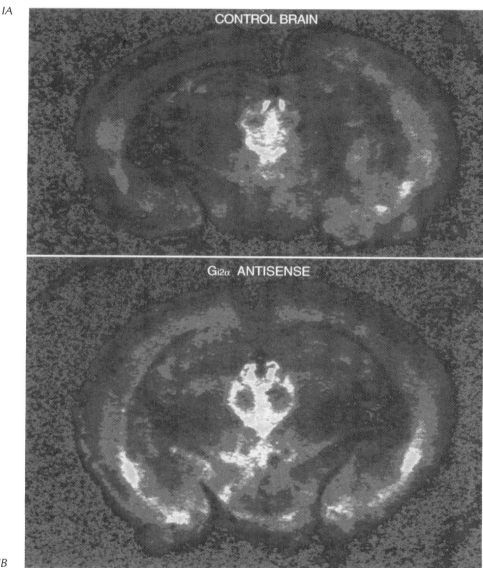

Fig. 5.8. Autoradiographic distribution of opioid μ receptors in mice previously injected i.c.v. either with vehicle (Panel IA, top; Panel IIA, facing page, top) or with antisense to $G_i2\alpha$ (Panel IB, bottom; Panel IIB, facing page, bottom), 24 hours before brains were taken for autoradiography. The brains were frozen in isopentane on dry ice and sectioned frozen at a thickness of 20 μm. The sections were thaw-mounted onto subbed slides and then stored at −80°C. Total binding (shown) was assessed with 5 nM [^{3}H]DAMGO in 50 mM Tris buffer (pH = 7.4) at 25°C

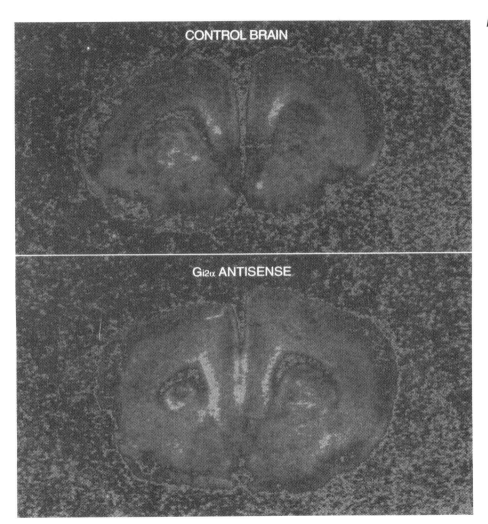

IIA

IIB

for 60 min. Nonspecific binding (not shown) was determined in the presence of 1 μM unlabelled DAMGO. Sections were subsequently washed 4 times in ice-cold 50 mM Tris (30 sec each), dipped once in distilled water, dried under a stream of dry air, then dessicated overnight at 4°C. [³H]Hyperfilm (Amersham) was exposed to the dried sections at −80°C for 8 weeks, after which film was developed in Kodak GBX developer for 5 min, fixed for 10 min, and washed in distilled water.

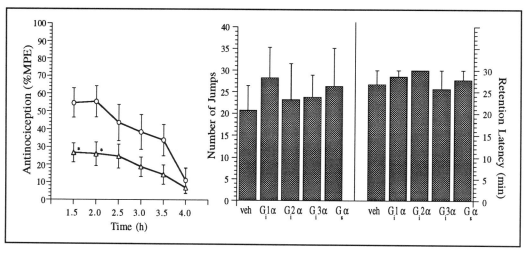

Fig. 5.9. (A) Tail-flick antinociception (%MPE ± S.E.M.) following i.c.v. morphine (10 μg). At 3 h prior to testing, the mice were injected i.c.v. with vehicle (veh) (circles) or $G_i2\alpha$ antisense (6 nmol) (triangles) (N = 14-20 per group). The asterisks denote significant difference (P < 0.05) from vehicle-treated mice. (B) Naloxone-precipitated jumps in 10 min (mean number ± S.E.M.) in the same animals as (A) and mean glass bead retention latency (min ± S.E.M.) in mice treated similarly to (A), except with 1 μg of morphine (N = 5 per group). Reprinted with permission from: Raffa RB, Goode TL, Martinez RP, Jacoby HI. Life Sci Pharmacol Letts 1996; 58:73-76.

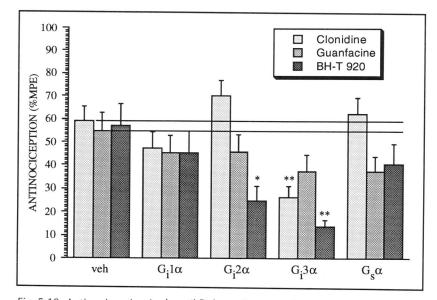

*Fig. 5.10. Antinociception in the tail-flick test (expressed as %MPE) following i.c.v. injection of the α_2-adrenoceptor agonists clonidine, guanfacine, or B-HT 920 to mice. At 18-24 h prior to the testing, the mice were injected i.c.v. with vehicle (veh) or antisense (6 nmol) to $G_i1\alpha$, $G_i2\alpha$, $G_i3\alpha$ or $G_s\alpha$. N = 14-30 mice per group. The Asterisks denote a significant difference (*P < 0.05; **P < 0.01) from vehicle-treated group. Reprinted with permission from: Raffa RB, Connelly CD, Chambers JR, Stone DJ. Life Sci/Pharmacol Letts 1996; 58:77-80.*

sense, mismatch or other comparators is common in work in vitro, but can be less compelling or prohibitively expensive in work in vivo. Where possible, the experimental design itself or a series of inter-related experiments, provides the proper controls.

With regard to the specific results of the studies described in this chapter, the lack of effect of pretreatment with antisense to $G_i 1\alpha$, $G_i 3\alpha$ or $G_s\alpha$ does not rule out the involvement of these subunits in the effects of μ agonists, it only suggests the relative sensitivities to the same dose of antisense. Likewise, the lack of effect of $G_s\alpha$ antisense, compared to the enhancement observed after anti-$G_s\alpha$ antibody pretreatment,[18] possibly reflects the fact that G_s mRNA is much more abundant than is G_i mRNA in brain, even though G_i protein is more abundant than G_s protein[30] — an important consideration in the interpretation of studies of this type. Likewise, some G-protein subunits not examined in these studies might also be involved in the transduction of antinociception.

CONCLUSIONS AND FUTURE DIRECTIONS

It has been a long-held belief that G-proteins, as ubiquitous transducers of the receptor-effector coupling mechanism of multiple receptor families in multiple cells, cannot confer any specificity of response—that receptors are the only site in the chain of events that would allow this specificity. The results of the series of studies presented in this chapter, we believe, suggest that this might not be the case. The results of these and other studies provide evidence that particular G-protein subunits preferentially mediate certain ligand-induced effects, whereas other subunits might mediate other effects (or 'side-effects') of the same ligand. Such differentiation implies that targeting specific G-protein subunits could produce precise biological effects. Such efforts could result in the discovery of important pharmacologic tools and drugs of enhanced therapeutic utility and safety.

ACKNOWLEDGMENTS

The present chapter summarizes the combined effort of many individuals, including James R. Chambers, Ellen E. Codd, Charlene D. Connelly, Tamara L. Goode, Alexander L. Kirifides, Rebecca P. Martinez, Dennis J. Stone, Jr. and Kenneth D. Wild. The editors wish to thank Jeanne Coughlin for her organizational, editorial and 'people' skills, without which this book would not have been possible.

REFERENCES

1. Wahlestedt C. Antisense oligonucleotide strategies in neuropharmacology. Trends Pharmacol Sci 1994; 15:42-46.
2. Bilsky EJ, Bernstein RN, Pasternak GW, Hruby VJ, Patel D, Lai J. Selective inhibition of [D-Ala^2,Glu^4]deltorphin antinociception by supraspinal, but not spinal, administration of an antisense oligodeoxynucleotide to an opioid delta receptor. Life Sci/Pharmacol Letts 1994; 55:PL37-43.

3. Mol JNM, van der Krol AR, eds. Antisense Nucleic Acids and Proteins: Fundamentals and Applications. New York: Marcel Dekker, Inc., 1991.
4. Chen X-H, Adams JU, Geller EB, DeRiel JK, Adler MW, Liu-Chen, L-Y. An antisense oligodeoxynucleotide to μ-opioid receptors inhibits μ-opioid receptor agonist-induced analgesia in rats. Eur J Pharmacol 1995; 275:105-108.
5. Tseng LF, Collins KA, Narita M, Kampine JP. The use of antisense oligodeoxynucleotides to block the spinal effects of κ_1 agonist-induced antinociception in the mouse. Analgesia 1995; 1:121-126.
6. Watson S, Arkinstall S, eds. The G-protein Linked Receptor Facts Book. London: Academic Press Inc., 1994.
7. Kurjan J, Taylor BL, eds. Signal Transduction: Prokaryotic and Simple Eukaryotic Systems. San Diego: Academic Press, Inc., 1993.
8. Birnbaumer L. G proteins in signal transduction. Annu Rev Pharmacol Toxicol 1990; 30:675-705.
9. Simon MI, Strathmann MP, Gautam N. Diversity of G proteins in signal transduction. Science 1991; 252:802-808.
10. Carr C, Loney C, Unson C, Knowler J, Milligan G. Chronic exposure of rat glioma C6 cells to cholera toxin induces loss of the α-subunit of the stimulatory guanine nucleotide-binding protein (Gs). Eur J Pharmacol 1990; 188:203-209.
11. Mitchell FM, Griffiths SL, Saggerson ED, Houslay MD, Knowler JT, Milligan G. Guanine-nucleotide-binding proteins expressed in rat white adipose tissue. Biochem J 1989; 262:403-408.
12. Jones DT, Reed RR. Molecular cloning of five GTP-binding protein cDNA species from rat olfactory neuroepithelium. J Biol Chem 1987; 262:14241-14249.
13. Sullivan KA, Liao Y-C, Alborzi A, Beiderman B, Chang F-H, Masters SB, Levinson AD, Bourne HR. Inhibitory and stimulatory G proteins of adenylate cyclase: cDNA and amino acid sequences of the α chains. Proc Natl Acad Sci 1986; 83:6687-6691.
14. Zigman JM, Westermark GT, LaMendola J, Steiner DF. Expression of cone transducin, $G_z\alpha$, and other G-protein α-subunit messenger ribonucleic acids in pancreatic islets. Endocrinol 1994; 135:31-37.
15. Janssen PAJ, Niemegeers CJE, Dony JGH. The inhibitory effect of fentanyl and other morphine-like analgesics on the warm water induced tail withdrawal reflex in rats. Arzn Forsch 1963; 13:502-507.
16. Hylden JLK, Wilcox GL. Intrathecal morphine in mice: a new technique. Eur J Pharmacol 1980; 67:313-316.
17. Garzón J. Minireview: cellular transduction regulated by μ- and δ-opioid receptors in supraspinal analgesia: GTP binding regulatory proteins as pharmacological targets. Analgesia 1995; 1:131-144.
18. Sánchez-Blázquez P, Garzón J. Intracerebroventricular injection of antibodies raised against $G_s\alpha$ enhances the supraspinal antinociception induced by morphine, β-endorphin and clonidine in mice. Life Sci/Pharmacol Letts 1992; 51:PL237-242.

19. Sánchez-Blázquez P, Juarros JL, Martinez-Peña Y, Castro MA, Garzón J. $G_{x/z}$ and G_{i2} transducer proteins on μ/δ opioid-mediated supraspinal antinociception. Life Sci/Pharmacol Letts 1993; 53:PL381-386.
20. Raffa RB, Martinez RP, Connelly CD. G-protein antisense oligodeoxyribonucleotides and μ-opioid supraspinal antinociception. Eur J Pharmacol 1994; 258:R5-7.
21. Mjanger E, Yaksh TL. Characteristics of dose-dependent antagonism by β-funaltrexamine of the antinociceptive effects of intrathecal mu agonists. J Pharmacol Exp Ther 1991; 258:544-550.
22. Stone Jr DJ, Wild KD, Raffa RB. Time-course of $G_i2α$ oligodeoxyribonucleotide antisense antagonism of μ-opioid antinociception in the mouse tail-immersion test. Analgesia 1995; 1:770-773.
23. Huidobro F, Maggiolo C. Some features of the abstinence syndrome to morphine in mice. Acta Physiol Latinoamer 1961; 11:201-209.
24. Kosersky DS, Harris RA, Harris LS. Naloxone-precipitated jumping activity in mice following the acute administration of morphine. Eur J Pharmacol 1974; 26:122-124.
25. Jacoby HI, Lopez I. A method for the evaluation of colonic propulsive motility in the mouse after i.c.v. administered compounds. Digest Dis Sci 1984; 29:551.
26. Raffa RB, Goode TL, Martinez RP, Jacoby HI. A $G_i2α$ antisense oligonucleotide differentiates morphine antinociception, constipation and acute dependence in mice. Life Sci/Pharmacol Letts 1996, 58:73-76.
27. Sánchez-Blázquez P, Garzón J. Cholera toxin and pertussis toxin on opioid and α-mediated supraspinal analgesia in mice. Life Sci 1991; 48: 1721-1727.
28. Raffa RB, Connelly CD, Chambers JR, Stone DJ. α-Subunit G-protein antisense oligodeoxynucleotide effects on supraspinal (i.c.v.) $α_2$-adrenoceptor antinociception in mice. Life Sci/Pharmacol Letts 1996; 58:77-80.
29. Kurose H, Regan JW, Caron MG, Lefkowitz RJ. Functional interactions of recombinant $α_2$ adrenergic receptor subtypes and G proteins in reconstituted phospholipid vesicles. Biochem 1991; 30:3335-3341.
30. Brann, MR, Collins RM, Spiegel A. Localization of mRNAs encoding the α-subunits of signal-transducing G-proteins within rat brain and among peripheral tissues. FEBS Letts 1987; 222:191-198.

CHAPTER 6

DOPAMINE ANTISENSE OLIGODEOXYNUCLEOTIDES AS POTENTIAL NOVEL TOOLS FOR STUDYING DRUG ABUSE

Benjamin Weiss, Long-Wu Zhou and Sui-Po Zhang

INTRODUCTION

Dopamine is involved in a number of neurological and psychiatric events,[1,2] among which is its role in the reinforcing effects of abused substances. For example, it has been suggested that the mesolimbic dopamine system is involved in the psychological dependence to several abused substances, including cocaine, amphetamine, ethanol, nicotine and opiates.[3-10] The mesolimbic dopamine system consists of dopaminergic cell bodies, originating in the ventral tegmental area and projecting to several brain regions, including the amygdaloid nucleus, olfactory tubercle and nucleus accumbens. The ventral tegmental-nucleus accumbens pathway appears to be involved in drug reinforcement and addiction, as well as in motivational states, stress responses and locomotor activity.[6,11] Further, it is believed that repeated exposure to some drugs produces adaptive changes in those brain areas which may be relevant to the reinforcement process.[4,12,13] For example, repeated administration of amphetamine, cocaine, and morphine enhances dopamine neurotransmission in the striatum, as well as in the nucleus accumbens, effects that may explain why repeated administration of these agents results in a sensitized motor behavior.[11,14]

Antisense Strategies for the Study of Receptor Mechanisms, edited by Robert B. Raffa and Frank Porreca. © 1996 R.G. Landes Company.

That the actions of cocaine may be mediated by dopamine receptors is supported by several additional lines of evidence. Cocaine produces an increase in the extracellular levels of dopamine in the mesolimbic areas of brain, an effect consistent with the well-known ability of cocaine to block the reuptake of dopamine into presynaptic nerve terminals of dopaminergic neurons. The density of postsynaptic dopamine receptors in the nucleus accumbens is also increased by repeated administration of cocaine, indicating a significant indirect effect of cocaine on dopaminergic neurons.[15,16] Dopamine uptake blockers, such as GBR 12909 and cocaine, decrease self-stimulation thresholds, while both D_1 and D_2 dopamine receptor antagonists (SCH 23390, sulpiride and pimozide) increase self-stimulation thresholds[17–19] and attenuate the reinforcing properties of psychomotor stimulants in animals.[20–22] Finally, several studies, using 6-hydroxydopamine as a dopaminergic neurotoxin, have shown that the destruction of dopaminergic neurons in the mesolimbic area blocks cocaine self-administration.[20,23]

Recently it has been suggested that the D_1 subtype of dopamine receptors may play a more important role in the reinforcing properties of cocaine than do D_2 receptors.[13,24] Using single cell electrophysiological recording and microiontophoretic techniques, Henry and White[25] found that chronic administration of cocaine produces a supersensitivity of D_1 dopamine receptors, but not D_2 dopamine receptors, in the nucleus accumbens; the neurons in the nucleus accumbens of cocaine-treated rats were sensitized to the inhibitory effects of the D_1 dopamine receptor agonist SKF 38393 but not to that of the D_2 agonist quinpirole. Similarly, the D_1 agonist SKF 81297 has been shown to function as a reinforcer in rhesus monkeys,[26] whereas administration of the D_1 receptor antagonist SCH 23390 into the ventromedial mesencephalon prevented the development of behavioral sensitization induced by repeated exposure to amphetamine or morphine.[11,27] Repeated administration of methamphetamine produced an upregulation in the density of D_1 receptors in the substantia nigra.[28] Finally, D_1 receptors appear to be involved in the lethal effects of cocaine in the mouse since pretreatment with SCH 23390 reduced the lethality of cocaine.[29]

Based on these data, in the past decade the search for pharmacotherapies to treat drug abuse has focused to a large extent on agents that alter dopaminergic function. However, despite intensive efforts, dopaminergic agents have not proven to be particularly effective for treating drug abuse. One possible reason is that the dopamine receptor system is far more complex than originally thought.

The recent cloning of the genes that encode the dopamine receptors revealed several subtypes of dopamine receptors in addition to the D_1 and D_2 dopamine receptors that have been known since 1988.[30,31] Studies of the structure of these receptors have shown that they likely belong to two major families: the D_1-like dopamine receptor family consisting of the D_1 and D_5 dopamine receptors and the D_2-like dopamine receptor family consisting of D_2, D_3 and D_4 dopamine receptors.[32-37] These receptor subtypes and their transcripts are unevenly distributed throughout the central nervous system, change at different rates during ontogeny[38] and aging,[34,39] and can be selectively regulated.[40-43]

Although great advances have been made in the molecular biology of the dopamine receptor subtypes, the specific biological functions of these subtypes of dopamine receptors are still unclear. Further, based on the recent molecular biological studies demonstrating the existence of multiple subtypes of dopamine receptors, new and serious questions have been raised about the selectivity of dopamine agonists and antagonists.[44,45] For example, despite its generally accepted specificity for D_1 dopamine receptors, SCH 23390 not only inhibits grooming induced by D_1 dopamine agonists, it also inhibits locomotor and stereotyped behaviors induced by D_2 dopamine agonists.[46] In addition, the dopamine receptor antagonist SCH 23390, which was thought to be specific for the D_1 receptors, has high affinities for both D_1 and D_5 dopamine receptors,[44] and quinpirole, which has been widely used as a prototypical D_2 dopamine receptor agonist, has been shown to have a higher affinity for D_3 receptors than for D_2 receptors.[45] Therefore, a new type of more selective agent to alter the function of the individual subtypes of the dopamine receptors must be developed. Oligodeoxynucleotides antisense to specific sequences of the mRNAs encoding the different dopamine receptor subtypes may provide such selective antagonists to alter the activity of these receptor subtypes and may provide a new and efficient means to study and perhaps treat certain forms of drug abuse.

ANTISENSE OLIGODEOXYNUCLEOTIDES—A NOVEL TOOL FOR INHIBITING THE EXPRESSION OF DOPAMINE RECEPTORS

Antisense oligodeoxynucleotides are short sequences of synthetic DNA that are complementary to a portion of a target mRNA. Through Watson-Crick interactions they hybridize with specific sequences of the transcripts encoding the receptor subtypes and thereby prevent translation of mRNA and the synthesis of the encoded protein.

In the past few years, antisense oligodeoxynucleotides have been used to interfere with the expression of a variety of genes in vitro[47-49] and in vivo.[50-52] A number of these studies have used antisense oligodeoxynucleotides in vivo to alter the expression of a variety of proteins in brain. These include c-fos,[53] the neuropeptide Y-Y1 receptor,[54] the N-methyl-D-aspartate receptor,[55] the progesterone receptor[56] and the dopamine receptor subtypes.[57-60] Taken together, these studies suggest that injection of antisense oligodeoxynucleotides targeted to different receptor mRNAs in brain can alter specific behaviors and the expression of specific proteins.

A major advantage of using an antisense strategy is the high specificity of the genetic code and the relatively simple, rational design and synthesis of oligodeoxynucleotides. Using antisense oligodeoxynucleotides to interfere with neurotransmission may have a further advantage of not inducing an upregulation of the receptors they are designed to inhibit.

In designing an antisense oligodeoxynucleotide, among the issues to be addressed are the size of the antisense, whether and how it should be modified, and which sequence on the mRNA should be targeted. To achieve a balance between ease of penetration into tissue and adequate hybridization to the selected target with a minimum interaction with nonselected targets, oligodeoxynucleotides are often in the range of about 20 nucleotides in length. Although the exact target site on the mRNA is still chosen largely by empirical means, many investigators have selected the sequence at or about the initiation (start) codon. Since natural oligodeoxynucleotides are rapidly degraded by endogenous nucleases,[61] they are often chemically modified to increase their stability, phosphorothioates being the derivative usually employed.

Because of the possibility of nonspecific effects of oligodeoxynucleotides, several controls should be used. One such control is to use random oligodeoxynucleotides which have the same base compositions but with nucleotides placed in a random sequence. Another test of specificity is to determine whether similar effects are produced by oligodeoxynucleotides targeted to different sites on the selected mRNA. Finally, the measured physiological or biochemical effects should be accompanied by specific changes in the levels of the target mRNAs and in the levels of translated proteins. These latter strictures may be modified somewhat since not all antisense oligodeoxynucleotide-RNA complexes are metabolized by RNase H, and therefore the levels of the transcripts may not necessarily be reduced even though their ability to translate proteins is impaired. Further, the levels of the translated proteins is often not reduced to the same level as is the biological function one is measuring. This may be explained by the fact that the reduction in the levels of a protein following antisense treatment is dependent upon its rate of degradation; slowly degraded proteins may not decrease substantially over a short period of time. Moreover, if the newly synthesized proteins are the ones that are functionally active, antisense oligodeoxynucleotides may, in fact, elicit a large biological effect while producing only a relatively small reduction in the target protein. We believe this to be the case in experiments in which antisense oligodeoxynucleotides targeted to the dopamine receptor transcripts are concerned.[62]

In our laboratory, 20-mer phosphorothioate-modified oligodeoxynucleotides targeted to D_1, D_2 and D_3 dopamine receptor mRNA were designed and synthesized. The D_1, D_2 and D_3 antisense oligodeoxynucleotides were targeted to the respective initiation codons of the mRNAs encoding the various subtypes of dopamine receptors.[57,59,63]

To determine the uptake and distribution of oligodeoxynucleotides in brain, we used a 20-mer phosphorothioated oligodeoxynucleotide complementary to the D_2 dopamine receptor mRNA, fluorescently labeled with fluorescein isothiocyanate, and injected into the lateral cerebral ventricles of mice. The results indicate that the oligodeoxynucleotide was rapidly taken up into the brain substance from the cerebral ventricles, and rapidly redistributed from the interstitial spaces and white matter to become highly concentrated in cell bodies.[64] This modified oligodeoxynucleotide was fairly stable in that there was relatively little degradation during the first 24 hr, but substantial degradation occurred over the next several days.[64]

The following sections briefly describe some studies on the application of antisense oligodeoxynucleotides directed to the D_1 and D_3 dopamine receptors as tools to uncover the specific functions of these subtypes of dopamine receptors. These subtypes of the dopamine receptors were chosen for discussion since these are the ones that more likely involve in the action of abused substances.

ADMINISTRATION OF D_1 ANTISENSE SPECIFICALLY INHIBITS D_1 DOPAMINE RECEPTOR-MEDIATED BEHAVIORS IN NORMAL MICE

To determine the behavioral consequences of administering a D_1 antisense oligodeoxynucleotide, mice were treated with intracerebroventricular injections of D_1 antisense, and the grooming behavior induced by subsequent challenge injections with the D_1 dopamine receptor agonist SKF 38393 was determined. Figure 6.1 shows that treatment with D_1 antisense caused a dose-related decrease in grooming behavior induced by SKF 38393; the ID_{50} value for the D_1 antisense was approximately 6 nmol/injection.

To determine the specificity by which the D_1 antisense inhibited D_1 dopamine receptor-mediated behavior in normal animals, mice were administered the D_1 antisense and then were challenged with high doses of SKF 38393 or with the D_2/D_3 dopamine receptor agonist quinpirole (Fig. 6.2). Figure 6.2A shows that intracerebroventricular administration of D_1 antisense produced a marked inhibition of grooming induced by SKF 38393, given at a dose that is 4 to 7 times its ED_{50} value.[65,66] By contrast, intracerebroventricular administration of D_1 antisense failed to significantly inhibit the stereotyped behavior induced by quinpirole, given at a dose that is 7 times its ED_{50} value[66] (Fig. 6.2B). These studies demonstrate the specificity with which D_1 antisense inhibits D_1 dopamine receptor-mediated behavior.

RECOVERY OF SKF38393 INDUCED-GROOMING BEHAVIOR AFTER CESSATION OF D_1 ANTISENSE TREATMENT

To determine whether the effect of D_1 antisense on SKF 38393-induced grooming behavior was reversible, mice were administered D_1

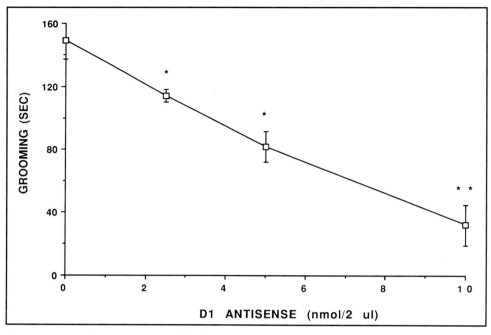

Fig. 6.1. Effect of D_1 antisense on SKF 38393-induced grooming behavior. Mice were administered three intracerebroventricular injections (at 12 hr intervals) of various doses of the D_1 antisense. Grooming behavior induced by acute challenge injections of the D_1 dopamine receptor agonist SKF 38393 (40 μmol/kg, s.c.) was measured 10 hr after the third injection of D_1 antisense. The values shown represent the increase in grooming score over that produced by acute challenge injections of saline. The control grooming score value was 40 sec. Each point represents the mean value from four mice. Vertical brackets indicate the S.E. *=$p<0.05$; **=$p<0.01$ compared with values obtained with no antisense treatment. Reprinted with permission: Zhang S-P, Zhou L-W, Weiss B. J Pharmacol Exp Ther 1994; 271:1462-1470.

antisense or vehicle repeatedly over a 5-day period, and grooming behavior induced by SKF 38393 was determined during treatment and after cessation of antisense or vehicle treatment. Figure 6.3 shows that a statistically significant reduction of SKF 38393-induced grooming behavior was observed when the mice were challenged with SKF 38393 from the 3rd to 6th day from the initiation of treatment with D_1 antisense. After cessation of treatment with D_1 antisense, a marked inhibition of the response to SKF 38393 persisted for more than 24 hr, then began to return to normal. By eight days after cessation of treatment with D_1 antisense, the effect of SKF 38393 was similar to that found in vehicle-treated animals. There was no significant alteration of SKF 38393-induced grooming behavior of any time point studied in vehicle-treated mice. These studies show that treatment with D_1 antisense produces a reversible inhibition of a D_1 receptor-mediated behavior, the halftime for recovery being about four days. This is

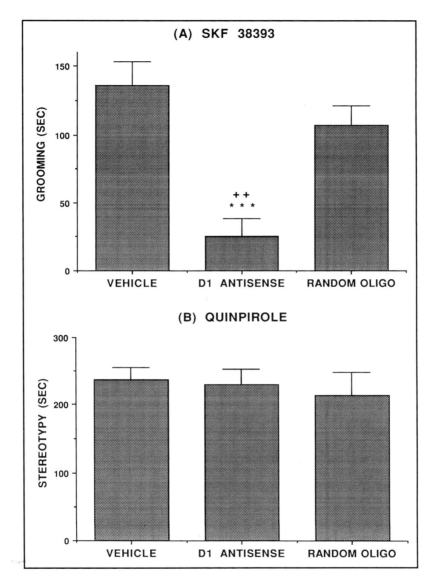

Fig. 6.2. Effect of D_1 antisense on grooming behavior induced by SKF 38393 and on stereotyped behavior induced by quinpirole. Mice were administered intracerebroventricular injections of vehicle (2 µl of artificial CSF), D_1 antisense (2.5 nmol/2 µl) or random oligo (2.5 nmol/2 µl) twice daily for six days. Grooming behavior (A) induced by acute injections of SKF 38393 (80 µmol/kg, s.c.) and stereotyped behavior (B) induced by acute injections of quinpirole (20 µmol/kg, s.c.) were measured 10 hr after the last injection of vehicle or oligomer. Each point represents the mean value from five mice. Vertical brackets indicate the S.E. *** = $p < 0.001$ compared with values from vehicle-treated mice. ++ = $p < 0.01$ compared with values from random oligo treated mice. Reprinted with permission: Zhang S-P, Zhou L-W, Weiss B. J Pharmacol Exp Ther 1994; 271:1462-1470.

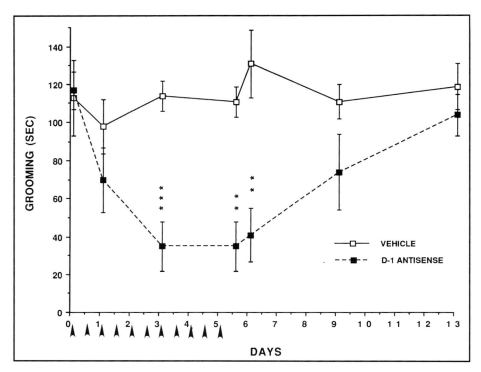

Fig. 6.3. Recovery of SKF 38393-induced grooming behavior after cessation of D_1 antisense treatment. Mice were administered intracerebroventricular injections of vehicle (2 µl of artificial CSF), or D_1 antisense (2.5 nmol/2 µl) at 12 hr intervals for a total of 11 injections (shown by the arrowheads). Grooming behavior induced by SKF 38393 (40 µmol/kg, s.c.) was measured 10 hr after the second and sixth injection of vehicle or D_1 antisense, and at 0.5, 1, 4 and 8 days after the last injection of vehicle or D_1 antisense. Each point represents the mean value from five mice. Vertical brackets indicate the S.E. * *= $p < 0.01$ and *** = $p < 0.001$ compared with values from vehicle-treated mice measured at the same time points. Reprinted with permission: Zhang S-P, Zhou L-W, Weiss B. J Pharmacol Exp Ther 1994; 271:1462-1470.

approximately the same time it takes D_1 dopamine receptors to recover following their irreversible inhibition.[62]

ADMINISTRATION OF D_1 ANTISENSE SPECIFICALLY INHIBITS ROTATIONAL BEHAVIOR INDUCED BY SKF 38393 IN 6-HYDROXYDOPAMINE-LESIONED MICE

To determine further the specificity by which D_1 antisense inhibited D_1 dopamine receptor-mediated behavior, we studied the effects of D_1 antisense in a model system in which animals were dopaminergically supersensitive; i.e., mice with unilateral lesions of the corpus striatum induced by the neurotoxin 6-hydroxydopamine. This model is particularly useful as the animals exhibit a rotational response to challenge injections to not only D_1 dopamine agonists (SKF 38393) but

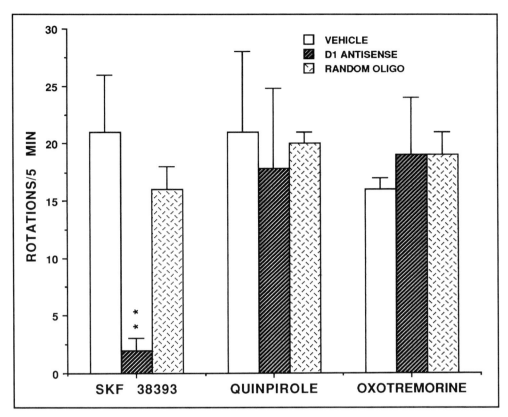

Fig. 6.4. D_1 antisense inhibits rotational behavior induced by SKF 38393 but not that induced by quinpirole or oxotremorine in 6-hydroxydopamine-lesioned mice. Mice with unilateral intrastriatal lesions induced by 6-hydroxydopamine were administered intracerebroventricular injections of vehicle (2 μl of artificial CSF), D_1 antisense (2.5 nmol/2 μl) or random oligo (2.5 nmol/2 μl) twice daily for 2 days. Mice were challenged with the D_1 dopamine receptor agonist SKF 38393 (40 μmol/kg, s.c.), the D_2 dopamine receptor agonist quinpirole (20 μmol/kg, s.c.) or the muscarinic cholinergic receptor agonist oxotremorine (5 μmol/kg, s.c.) 10 hr after the last injection of vehicle or oligomer, and rotational behavior was assessed. The rotational score in response to challenge injections of vehicle was zero. Each point represents the mean value from three to four mice. Vertical brackets indicate the S.E. ** = $p < 0.01$ compared with values from vehicle- or random oligo-treated mice. Reprinted with permission: Zhang S-P, Zhou L-W, Weiss B. J Pharmacol Exp Ther 1994; 271:1462-1470.

also to that of D_2 dopamine agonists (quinpirole)[67] and muscarinic cholinergic agonists (oxotremorine).[68] Accordingly, mice with unilateral intrastriatal 6-hydroxydopamine lesions were administered intracerebroventricular injections of D_1 antisense, and then challenged with SKF 38393, quinpirole and oxotremorine. Figure 6.4 shows that treatment of mice with D_1 antisense almost completely inhibited rotational behavior induced by SKF 38393, but did not significantly inhibit rotations induced by quinpirole or oxotremorine. Similar treatment of mice with the random oligomer failed to alter the rotational response to any of the agonists studied.

ADMINISTRATION OF D_1 ANTISENSE DECREASES D_1 DOPAMINE RECEPTORS IN MOUSE BRAIN

To determine whether the intracerebroventricular administration of D_1 antisense inhibited the function of D_1 dopamine receptor mRNA (i.e., reduced the formation of D_1 receptor protein), the levels of D_1 dopamine receptors in brains of mice treated with D_1 antisense were analyzed. Mice were continuously infused with D_1 antisense or vehicle into the lateral cerebral ventricles using Alzet osmotic minipump brain infusion kits. After five days of infusion, the brains were removed, sectioned and D_1 dopamine receptors determined using autoradiographic techniques (as assessed by the binding of [^{3}H] SCH 23390). Figure 6.5 shows that continuous infusion of D_1 antisense caused a reduction in the density of D_1 dopamine receptors in mouse corpus striatum, nucleus accumbens and olfactory tubercle.

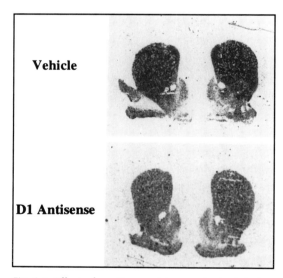

Fig. 6.5. Effect of continuous intraventricular infusion of D_1 antisense on D_1 dopamine receptors in mouse brain. Using Alzet brain infusion kits, in which osmotic minipumps were implanted subcutaneously onto the backs of mice, animals were continuously infused with D_1 antisense (1.5 nmol in 1 μl/hr) or vehicle (artificial CSF, 1 μl/hr) into the lateral cerebral ventricles. After five days of infusion, the brains were removed, sectioned, and receptor autoradiography performed using the D_1 dopamine receptor ligand [^{3}H]-SCH 23390 (2 nM). The results show that continuous infusion of D_1 antisense caused a reduction in the levels of D_1 dopamine receptors in mouse corpus striatum, nucleus accumbens and olfactory tubercle.

ADMINISTRATION OF D_3 ANTISENSE INCREASES LOCOMOTOR BEHAVIOR INDUCED BY CONTINUOUSLY INFUSING QUINPIROLE IN MICE

Although there is a large body of evidence demonstrating that D_1 dopamine receptors mediate grooming behavior in rodents,[65,69] there is relatively little data concerning the function of the D_3 dopamine receptor subtype. Nevertheless, the bulk of the available evidence suggests that D_3 dopamine receptors mediate locomotor activity.[70] Therefore, we used this behavior as one measure of D_3 dopamine receptor activity and examined the effect on locomotion of treating mice with a D_3 antisense oligodeoxynucleotide. Like the other dopamine antisense oligodeoxynucleotides, this antisense was also targeted near the initiation codon, in this case to nucleotides +3 to +22 of the D_3 dopamine receptor mRNA.

Continuous infusion of quinpirole (a D_2/D_3 dopamine agonist), using a subcutaneously implanted Alzet minipump, produces significant locomotor behavior within two days after implanting the pump. This locomotor behavior, which is not blocked by acute injections of the D_2 dopamine antagonist sulpiride, continues for the 6 days of infusion.[66] We hypothesized that the locomotor behavior induced by continuously infusing quinpirole might have resulted from an alteration of D_3 dopamine receptors in the nucleus accumbens. To determine whether the D_3 dopamine receptor was involved in this locomotor behavior, mice were administered intracerebroventricular injections of D_1, D_2 and D_3 antisense, vehicle or random oligomer every 12 hr for 7 days. After the fifth of these injections, the mice were implanted with Alzet osmotic minipumps containing either quinpirole or vehicle. Locomotor behavior was measured at 1, 2, 3 and 4 days after implanting the pumps.

Figure 6.6 shows that treating mice with D_3 antisense, but not with D_1 and D_2 antisense, significantly increased the locomotor behavior seen at 2, 3 and 4 days after continuously infusing quinpirole. The results suggest that locomotor behavior induced by infusing quinpirole might be due to the down-regulation of D_3 dopamine receptors, and that D_3 antisense may potentiate this decreased function of the D_3 dopamine receptors by reducing their synthesis.

POSSIBLE MECHANISMS BY WHICH DOPAMINE RECEPTOR ANTISENSE OLIGODEOXYNUCLEOTIDES MAY BE USED TO STUDY OR TREAT DRUG ABUSE

The specificity with which antisense oligodeoxynucleotides act suggests that these novel agents might be used to study further the dopaminergic mechanisms involved in certain forms of drug abuse. For example, these results suggest that the administration of D_1 antisense into certain brain areas, by inhibiting the expression of D_1 dopamine receptors and D_1 dopamine receptor-mediated behaviors, might be helpful

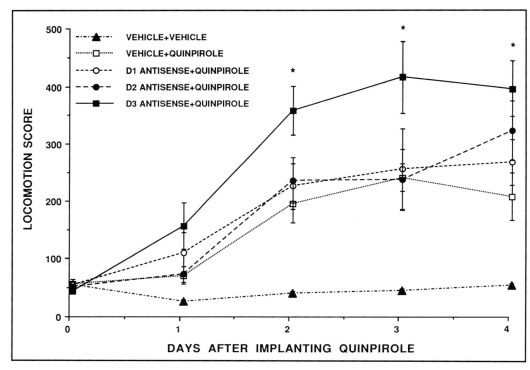

Fig. 6.6. D_3 antisense increases locomotor behavior induced by continuously infusing quinpirole in mice. Mice were administered intracerebroventricular injections of D_1, D_2 or D_3 antisense (2.5 nmol/2 μl), vehicle or random oligomer twice daily for seven days. After five such injections, mice were implanted subcutaneously with Alzet osmotic minipumps designed to deliver quinpirole at a rate of 2.5 μmol/kg/hr. Control mice were implanted with minipumps containing the vehicle (12% ascorbic acid and 50% dimethylsulfoxide). Locomotor behavior was assessed between 9:00 A.M. and 12:00 P.M. The mice were transferred to a behavioral testing room, placed in individual cages, and allowed to adapt to the new environment for 20 min before observing the behavior. Each point represents the mean locomotor score from five mice. Vertical brackets indicate the standard error. * = $p < 0.05$ compared with values obtained from mice treated with vehicle or quinpirole alone at the same time points.

for decreasing the high dopaminergic neurotransmission induced by some commonly abused drugs such as cocaine[71] and amphetamine.[9] This possibility is supported by the results showing that cocaine increases the expression of c-fos and alters a number of other biochemical events mediated by activation of D_1 dopamine receptors.[72]

The induction of the immediate-early gene c-fos has become widely accepted as a marker of neuronal activation.[73] In support of the mechanism that c-fos is involved in the actions of certain abused substances that interact with the dopamine system are the reports of Heilig et al[74] who showed that bilateral administration of a c-fos antisense into the nucleus accumbens blocked cocaine-induced locomotor activity, and that of Dragunow et al[75] who found that the unilateral infusion of c-fos antisense into the striatum prevents amphetamine-induced expression of c-fos and leads to amphetamine-induced ipsilateral rotational behavior. Further, administration of cocaine and amphetamine increases c-fos expression in rat striatum and nucleus accumbens.[13,76] This activation of c-fos has been reported to be regulated by cyclic AMP-dependent kinase, which is increased by D_1 dopamine receptor stimulation.[77,78]

Based on these data, a possible action of D_1 antisense in the treatment of drug abuse may be depicted as follows (see Fig. 6.7): administration of cocaine or amphetamine increases the interaction of dopamine with D_1 dopamine receptors in nucleus accumbens and corpus striatum by blocking the reuptake of dopamine into presynaptic nerve terminals or by stimulating the release of dopamine. This results in increased activities of adenylate cyclase and cyclic AMP-dependent protein

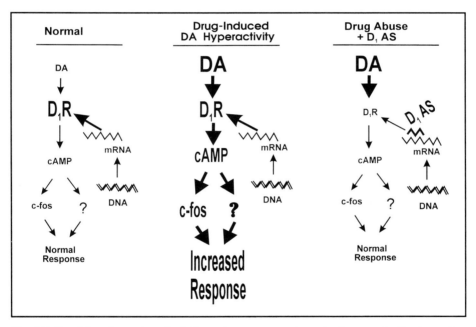

Fig. 6.7. Possible scheme by which D_1 antisense may reduce the increase in D_1 dopamine receptor-mediated responses to abused substances. In a normal cell, dopamine, through its interaction with postsynaptic D_1 dopamine receptors, activates adenylate cyclase, resulting in an increased intracellular concentration of cyclic AMP. This, in turn, through mechanisms, such as that involving cyclic AMP-dependent protein kinase and increases in the levels of c-fos, results in a characteristic dopaminergic response. Administration of certain types of abused substances (e.g., cocaine, amphetamine), causes an elevation in the concentration of dopamine at dopaminergic synapses, resulting in a cascade of events leading to an increase in dopaminergic behavior mediated by D_1 dopamine receptors. Treatment with D_1 antisense, by reducing the synthesis of postsynaptic D_1 dopamine receptors, causes a reduction in the functional pool of these receptors. In this scenario, this would lead to a restoration to normal of the D_1 dopamine receptor-mediated behavior despite the higher concentration of dopamine in the synapse.

kinase, and increases in the expression of certain proteins such as c-fos. All these changes lead to marked increases in D_1 dopamine receptor-mediated functional responses. Administration of D_1 antisense, by inhibiting the translation of D_1 dopamine receptor mRNA and the synthesis of D_1 dopamine receptors, decreases the functional pool of D_1 dopamine receptors,[62] thereby attenuating the abnormal responses induced by these agents.

There is also evidence suggesting that D_3 dopamine receptors are involved in drug abuse. It has been reported that D_3 dopamine receptor mRNA is expressed predominantly in ventral striatum (limbic dopamine projection areas), substantia nigra and nucleus accumbens. Moreover, some relatively selective D_3 antagonists, such as (+)-AJ 76, (+)-UH 232 and (-)-DS 121, which have a high affinity for the D_3 dopamine receptors at presynaptic and postsynaptic sites in these areas,[79,80] were effective at reducing the reinforcing effects induced by cocaine.[81,82] On the other hand, Caine and Koob[83] reported that the D_3 dopamine agonists, 7-hydroxy-N,N-di-n-propyl-2-aminotetralin (7-OH DPAT) and quinpirole, also decreased cocaine self-administration in the rat. This action apparently is on presynaptic D_3 dopamine autoreceptors, which when activated would cause a decreased release of dopamine.

Other studies indicated that the levels of D_3 dopamine receptors increase after chronic exposure to certain abused substances, although these changes were not consistently seen. Segal et al[84] found a marked elevation of the number of D_3 dopamine receptors and of the expression of D_3 dopamine receptors in the nucleus accumbens, caudate and putamen of human brain in cocaine addicted individuals. However, in rodents, chronic cocaine treatment significantly increased the levels of D_3 dopamine receptors in striatum but decreased the levels in nucleus accumbens.[85]

In our studies, we found that D_3 antisense, but not D_1 or D_2 antisense, increased locomotor behavior induced by continuously infusing the D_2/D_3 dopamine agonist quinpirole. These results are consistent with the evidence that activation of D_3 dopamine receptors inhibits locomotor activity[80] and that relatively selective D_3 antagonists increase locomotor activity.[70]

In addition to the D_1 and D_2 dopamine receptors, D_2 autoreceptors may also play an important role in controlling the release of dopamine induced by cocaine.[82] Low doses of the mixed D_1/D_2 agonist apomorphine and D_2 dopamine agonists, such as BHT 920, which bind with a high affinity to D_2 autoreceptors, decrease the release of dopamine and decrease locomotor activity.[86,87] Another more recent study[88] showed that D_2 antisense administered unilaterally into the substantia nigra caused a decrease in the density D_2 autoreceptors in the nigra, an increased in the release of dopamine from striatal slices, and a marked contralateral rotational behavior in response to cocaine.

This study provides evidence consistent with the hypothesis that the inhibition of D_2 dopamine receptors is involved in cocaine-induced motor behavior, and suggests further that dopamine receptor antisense oligodeoxynucleotides may be useful for studying the mechanisms by which abused substances act.

Although many questions remain unanswered regarding the specific dopaminergic mechanisms that may be involved in drug abuse, advances in the molecular biology of the dopamine receptor subtypes and the recent studies showing that antisense oligodeoxynucleotides have selective effects on behaviors mediated by the different subtypes of dopamine receptors, suggest that antisense oligodeoxynucleotides may provide a novel and potentially useful tool for uncovering the biological function of dopamine receptor subtypes in specific brain areas, and for understanding further the biochemical and molecular mechanisms involved in drug abuse. They also suggest the possibility of developing a new molecular strategy for treating drug abuse.

SUMMARY

Dopamine, through its interaction with one of several different subtypes of dopamine receptors, appears to play an important role in the effects of a number of widely abused substances, including cocaine, amphetamine, morphine and alcohol. Although several strategies have evolved using antidopaminergic agents to study the mechanisms involved in the drug reinforcement effect of brain stimulation, the currently available drugs that influence the dopaminergic system are relatively nonselective in their actions and have, therefore, enjoyed limited success in treating drug abuse. Herein we present evidence that antisense oligodeoxynucleotides directed to the mRNAs encoding the different dopamine receptor subtypes produce selective alterations in behavioral responses mediated by these receptors. Particular attention is focused on the D_1 and D_3 dopamine receptor subtypes as these are the ones most likely to be involved in drug abuse. The results show that intracerebroventricular administration of a D_1 antisense oligodeoxynucleotide to mice produces a dose-related, reversible inhibition in the behavioral response to a D_1 dopamine receptor agonist. This effect was seen in both normal and in dopaminergically supersensitive mice and was selective in that there was no significant change in the behavioral response to a D_2 dopamine receptor agonist or to a muscarinic cholinergic agonist. The D_1 antisense also caused a reduction in the levels of D_1 dopamine receptors in corpus striatum, nucleus accumbens and olfactory tubercle, areas that are thought to be involved in some of the actions of abused substances. Intracerebroventricular administration of a D_3 antisense oligodeoxynucleotide caused an increase in locomotor behavior induced by the continuous infusion of quinpirole in mice. No such effects were seen upon administration of similar concentrations of a D_1 or D_2 antisense oligodeoxynucleotide.

These results suggest that antisense oligodeoxynucleotides directed to the mRNAs encoding specific dopamine receptor subtypes may be useful for studying the mechanisms by which abused substances act and perhaps suggest novel pharmacotherapies for treating drug abuse.

Acknowledgments

This work is supported by a grant from the National Institute of Mental Health MH42148. We thank LYNX Therapeutics for their generous support in providing some of the antisense oligodeoxynucleotides used in this study.

References

1. Seeman P, Bzowej NH, Guan H-C et al. Human brain D_1 and D_2 dopamine receptors in schizophrenia, Alzheimer's, Parkinson's, and Huntington's diseases. Neuropsychopharmacology 1987; 1:5-15.
2. Carlsson A. The current status of the dopamine hypothesis of schizophrenia. Neuropsychopharmacology 1988; 1:179-186.
3. Bozarth MA. New perspectives on cocaine addiction: recent findings from animal research. Can J Physiol Pharmacol 1989; 67:1158-1167.
4. Wise RA, Rompre P-P. Brain dopamine and reward. Annu Rev Psychol 1989; 40:191-225.
5. Kuhar MJ, Ritz MC, Boja JW. The dopamine hypothesis of the reinforcing properties of cocaine. Trends Neurosci 1991; 14:299-302.
6. Fibiger HC, Phillips AG, Brown EE. The neurobiology of cocaine-induced reinforcement. Ciba Foundation Symposium 1992; 166:96-111.
7. Koob GF, Maldonado R, Stinus L. Neural substrates of opiate withdrawal. Trends Neurosci 1992; 15:186-191.
8. Toth E, Sershen H, Hashim A et al. Effect of nicotine on extracellular levels of neurotransmitters assessed by microdialysis in various brain regions: role of glutamic acid. Neurochem Res 1992; 17:265-271.
9. McMillen BA. CNS stimulants: two distinct mechanisms of action for amphetamine-like drugs. Trends Pharmacol Sci 1983; 4:429-432.
10. Gorwood P, Ades J, Feingold J. Are genes coding for dopamine receptors implicated in alcoholism? Eur Psychiatry 1994; 9:63-69.
11. Kalivas PW, Stewart J. Dopamine transmission in the initiation and expression of drug- and stress-induced sensitization of motor activity. Brain Res Rev 1991; 16:223-244.
12. Nestler EJ, Hope BT, Widnell KL. Drug addiction: a model for the molecular basis of neural plasticity. Neuron 1993; 11:995-1006.
13. Koob GF. Drugs of abuse: anatomy, pharmacology and function of reward pathways. Trends Pharmacol Sci 1992; 13:177-184.
14. Robinson TE, Jurson PA, Bennett JA et al. Persistent sensitization of dopamine neurotransmission in ventral striatum (nucleus accumbens) produced by prior experience with (+)-amphetamine: a microdialysis study in freely moving rats. Brain Res 1988; 462:211-222.
15. Goeders NE, Kuhar MJ. Chronic cocaine administration induces oppo-

site changes in dopamine receptors in the striatum and nucleus accumbens. Alcohol Drug Res 1987; 7:207-216.
16. Peris J, Boyson SJ, Cass WA et al. Persistence of neurochemical changes in dopamine systems after repeated cocaine administration. J Pharmacol Exp Ther 1990; 253:38-44.
17. Fibiger HC, Phillips AG. Mesocorticolimbic dopamine systems and reward. Ann NY Acad Sci 1988; 537:206-215.
18. Nakajima S, McKenzie GM. Reduction of the rewarding effect of brain stimulation by a blockade of dopamine D_1 receptor with SCH 23390. Pharmacol Biochem Behav 1986; 24:919-923.
19. Rompre P-P, Bauco P. GBR 12909 reverses the SCH 23390 inhibition of rewarding effects of brain stimulation. Eur J Pharmacol 1990; 182:181-184.
20. Goeders NE, Smith JE. Reinforcing properties of cocaine in the medial prefrontal cortex: primary action on presynaptic dopaminergic terminals. Pharmacol Biochem Behav 1986; 25:191-199.
21. Ettenberg A, Pettit HO, Bloom FE et al. Heroin and cocaine intravenous self-administration in rats: mediation by separate neural systems. Psychopharmacol 1982; 78:204-209.
22. De Wit H, Wise RA. Blockade of cocaine reinforcement in rats with the dopamine receptor blocker pimozide, but not with the noradrenergic blockers phentolamine or phenoxybenzamine. Can J Psychol 1977; 31:195-203.
23. Roberts DCS. Breaking pionts on a progressive ratio schedule reinforced by intravenous apomorphine increase daily following 6-hydroxydopamine lesions of the nucleus accumbens. Pharmacol Biochem Behav 1989; 32:43-47.
24. Robledo P, Maldonado-Lopez R, Koob GF. Role of dopamine receptors in the nucleus accumbens in the rewarding properties of cocaine. Ann NY Acad Sci 1992; 654:509-512.
25. Henry DJ, White FJ. Repeated cocaine administration causes persistent enhancement of D_1 dopamine receptor sensitivity within the rat nucleus accumbens. J Pharmacol Exp Ther 1991; 258:882-890.
26. Weed MR, Woolverton WL. Self-administration of cocaine and dopamine D_1 agonists under a progressive-ratio schedule in rhesus monkeys. 57th Annual Scientific Meeting: College on Problems of Drug Dependence 1995; 57:151.
27. Stewart J, Vezina P. Microinjections of SCH-23390 into the vental tegmental area and substantia nigra pars reticulata attenuate the development of sensitization to the locomotor activating effects of systemic amphetamine. Brain Res 1989; 495:401-406.
28. Ujike H, Akiyama K, Nishikawa H et al. Lasting increase in D_1 dopamine receptors in the lateral part of the substantia nigra pars reticulata after subchronic methamphetamine administration. Brain Res 1991; 540: 159-163.
29. Schechter MD, Meehan SM. Role of dopamine D_1 receptors in cocaine

lethality. Pharmacol Biochem Behav 1995; 51:521-523.
30. Spano PF, Govoni S, Trabucci M. Studies on the pharmacological properties of dopamine receptors in various areas of the central nervous system. Adv Biochem Psychopharmacol 1978; 19:155-165.
31. Kebabian JW, Calne DB. Multiple receptors for dopamine. Nature 1979; 277:93-96.
32. Civelli O, Bunzow JR, Grandy DK et al. Molecular biology of the dopamine receptors. Eur J Pharmacol Mol Pharmacol 1991; 207:277-286.
33. Seeman P. Dopamine receptor sequences. Therapeutic levels of neuroleptics occupy D_2 receptors, clozapine occupies D_4. Neuropsychopharmacology 1992; 7:261-284.
34. Weiss B, Chen JF, Zhang S et al. Developmental and age-related changes in the D_2 dopamine receptor mRNA subtypes in rat brain. Neurochem Int 1992; 20 Suppl:49S-58S.
35. Gingrich JA, Caron MG. Recent advances in the molecular biology of dopamine receptors. Annu Rev Neurosci 1993; 16:299-321.
36. Sibley DR, Monsma Jr FJ. Molecular biology of dopamine receptors. Trends Pharmacol Sci 1992; 13:61-69.
37. Sokoloff P, Schwartz J-C. Novel dopamine receptors half a decade later. Trends Pharmacol Sci 1995; 16:270-275.
38. Chen JF, Weiss B. Ontogenetic expression of D_2 dopamine receptor mRNA in rat corpus striatum. Dev Brain Res 1991; 63:95-104.
39. Mesco ER, Carlson SG, Joseph JA et al. Decreased striatal D_2 dopamine receptor mRNA synthesis during aging. Mol Brain Res 1993; 17:160-162.
40. Chen JF, Aloyo VJ, Weiss B. Continuous treatment with the D_2 dopamine receptor agonist quinpirole decreases D_2 dopamine receptors, D_2 dopamine receptor messenger RNA and proenkephalin messenger RNA, and increases mu opioid receptors in mouse striatum. Neuroscience 1993; 54:669-680.
41. Weiss B, Zhou L-W, Chen JF et al. Distribution and modulation of the D_2 dopamine receptor mRNA in mouse brain: molecular and behavioral correlates. Adv Biosci 1990; 77:9-25.
42. Buckland PR, O'Donovan MC, McGuffin P. Changes in dopamine D_1, D_2 and D_3 receptor mRNA levels in rat brain following antipsychotic treatment. Psychopharmacology 1992; 106:479-483.
43. Srivastava LK, Morency MA, Bajwa SB et al. Effect of haloperidol on expression of dopamine D2 receptor mRNAs in rat brain. J Mol Neurosci 1990; 2:155-161.
44. Sunahara RK, Guan HC, O'Dowd BF et al. Cloning of the gene for a human dopamine D5 receptor with higher affinity for dopamine than D1. Nature 1991; 350:614-619.
45. Sokoloff P, Giros B, Martres MP et al. Molecular cloning and characterization of a novel dopamine receptor (D3) as a target for neuroleptics. Nature 1990; 347:146-151.
46. Breese GR, Mueller RA. SCH-23390 antagonism of a D-2 dopamine ago-

nist depends upon catecholaminergic neurons. Eur J Pharmacol 1985; 113:109-114.
47. Stein CA, Cheng Y-C. Antisense oligonucleotides as therapeutic agents—is the bullet really magical. Science 1993; 261:1004-1012.
48. Helene C. Rational design of sequence-specific oncogene inhibitors based on antisense and antigene oligonucleotides. Eur J Cancer 1991; 27:1466-1471.
49. Crooke ST. Therapeutic applications of oligonucleotides. Annu Rev Pharmacol Toxicol 1992; 32:329-376.
50. Pilowsky PM, Suzuki S, Minson JB. Antisense oligonucleotides: a new tool in neuroscience. Clin Exp Pharmacol Physiol 1994; 21:935-944.
51. Simantov R. Neurotransporters: regulation, involvement in neurotoxicity, and the usefulness of antisense nucleic acids. Biochem Pharmacol 1995; 50:435-442.
52. Zon G. Brief overview of control of genetic expression by antisense oligonucleotides and in vivo applications. Mol Neurobiol 1995; 10:219-229.
53. Chiasson BJ, Hooper ML, Murphy PR et al. Antisense oligonucleotide eliminates in vivo expression of c-*fos* in mammalian brain. Eur J Pharmacol Mol Pharmacol 1992; 227:451-453.
54. Wahlestedt C, Pich EM, Koob GF et al. Modulation of anxiety and neuropeptide Y-Y1 receptors by antisense oligodeoxynucleotides. Science 1993; 259:528-531.
55. Wahlestedt C, Golanov E, Yamamoto S et al. Antisense oligodeoxynucleotides to NMDA-R1 receptor channel protect cortical neurons from excitotoxicity and reduce focal ischaemic infarctions. Nature 1993; 363:260-263.
56. Pollio G, Xue P, Zanisi M et al. Antisense oligonucleotide blocks progesterone-induced lordosis behavior in ovariectomized rats. Molecular Brain Res 1993; 19:135-139.
57. Weiss B, Zhou L-W, Zhang S-P et al. Antisense oligodeoxynucleotide inhibits D_2 dopamine receptor-mediated behavior and D_2 messenger RNA. Neuroscience 1993; 55:607-612.
58. Zhang M, Creese I. Antisense oligodeoxynucleotide reduces brain dopamine D_2 receptors: Behavioral correlates. Neurosci Lett 1993; 161:223-226.
59. Zhou L-W, Zhang S-P, Qin Z-H et al. In vivo administration of an oligodeoxynucleotide antisense to the D_2 dopamine receptor mRNA inhibits D_2 dopamine receptor-mediated behavior and the expression of D_2 dopamine receptors in mouse striatum. J Pharmacol Exp Ther 1994; 268:1015-1023.
60. Nissbrandt H, Ekman A, Eriksson E et al. Dopamine D3 receptor antisense influences dopamine synthesis in rat brain. Neuroreport 1995; 6:573-576.
61. Ott J, Eckstein F. Protection of oligonucleotide primers against degradation by DNA polymerase I. Biochemistry 1987; 26:8237-8241.
62. Qin Z-H, Zhou L-W, Zhang S-P et al. D_2 dopamine receptor antisense oligodeoxynucleotide inhibits the synthesis of a functional pool of D_2 dopamine receptors. Mol Pharmacol 1995; 48:730-737.

63. Zhang S-P, Zhou L-W, Weiss B. Oligodeoxynucleotide antisense to the D_1 dopamine receptor mRNA inhibits D_1 dopamine receptor-mediated behaviors in normal mice and in mice lesioned with 6-hydroxydopamine. J Pharmacol Exp Ther 1994; 271:1462-1470.
64. Zhang S-P, Zhou L-W, Morabito, M. et al. Uptake and distribution of fluorescein-labeled D_2 dopamine antisense oligodeoxynucleotide in mouse brain. J Mol Neurosci 1996; in press.
65. Starr BS, Starr MS. Differential effects of dopamine D_1 and D_2 agonists and antagonists on velocity of movement, rearing and grooming in the mouse. Neuropharmacology 1986; 25:455-463.
66. Zhou LW, Qin ZH, Weiss B. Downregulation of stereotyped behavior and production of latent locomotor behaviors in mice treated continuously with quinpirole. Neuropsychopharmacology 1991; 4:47-55.
67. Winkler JD, Weiss B. Effect of continuous exposure to selective D1 and D2 dopaminergic agonists on rotational behavior in supersensitive mice. J Pharmacol Exp Ther 1989; 249:507-516.
68. Zhou L-W, Zhang S-P, Connell TA et al. AF64A lesions of mouse striatum result in ipsilateral rotations to D_2 dopamine agonists but contralateral rotations to muscarinic cholinergic agonists. J Pharmacol Exp Ther 1993; 264:824-830.
69. Molloy AG, Waddington JL. Dopaminergic behavior stereospecifically promoted by the D_1 agonist R-SKF 38393 and selectively blocked by the D_1 antagonist SCH 23390. Psychopharmacology 1984; 82:409-410.
70. Svensson KA, Waters N, Sonesson C. (-)-DS121, a novel dopamine D_3 and autoreceptor preferring antagonist: effects on locomotor activity in the rat. Soc Neurosci 1993; 19:89.
71. Hurd YL, Weiss F, Koob GF et al. Cocaine reinforcement and extracellular dopamine overflow in rat nucleus accumbens: an in vivo microdialysis study. Brain Res 1989; 498:199-203.
72. Terwilliger RZ, Beitner-Johnson D, Sevarino KA et al. A general role for adaptations in G-proteins and the cyclic AMP system in mediating the chronic actions of morphine and cocaine on neuronal function. Brain Res 1991; 548:100-110.
73. Morgan JI, Curran T. Stimulus-transcription coupling the nervous system: involvement of the inducible proto-oncogenes fos and jun. Annu Rev Neurosci 1991; 14:421-451.
74. Heilig M, Engel JA, Soderpalm B. C-fos antisense in the nucleus accumbens blocks the locomotor stimulant action of cocaine. Eur J Pharmacol 1993; 236:339-340.
75. Dragunow M, Lawlor P, Chiasson B et al. c-*fos* antisense generates apomorphine and amphetamine-induced rotation. Neuroreport 1993; 5:305-306.
76. Young ST, Porrino LJ, Iadarola MJ. Cocaine induces striatal c-Fos-immunoreactive proteins via dopaminergic D_1 receptors. Proc Natl Acad Sci USA 1991; 88:1291-1295.

77. Graybiel AM, Moratalla R, Robertson HA. Amphetamine and cocaine induce drug-specific activation of the c-fos gene in striosome-matrix compartments and limbic subdivision of the striatum. Proc Natl Acad Sci USA 1990; 87:6912-6916.
78. Robertson GS, Vincent SR, Fibiger HC. Striatonigral projection neurons contain D1 dopamine receptor-activated c-fos. Brain Res 1990; 523:288-290.
79. Cao G-H, Woolverton WL. Antagonism of quinpirole by (+)-AJ 76: possible involvement of D_3 receptors. Eur J Pharmacol 1991; 209:285-286.
80. Waters N, Svensson K, Haadsma-Svensson SR et al. The dopamine D_3-receptor: a postsynaptic receptor inhibitory on rat locomotor activity. J Neural Transm Gen Sect 1993; 94:9-11.
81. Richardson NR, Piercey MF, Svensson K et al. Antagonism of cocaine self-administration by the preferential dopamine autoreceptor antagonist, (+)-AJ 76. Brain Res 1993; 619:15-21.
82. Roberts DCS, Ranaldi R. Effect of dopaminergic drugs on cocaine reinforcement. Clinical Neuropharmacol 1995; 18:S84-S95.
83. Caine SB, Koob GF. Modulation of cocaine self-administration in the rat through D-3 dopamine receptors. Science 1993; 260:1814-1816.
84. Segal D, Staley JK, Basile M et al. Neuroadaptive regulation of the D_3 dopamine receptor by cocaine. 57th Annual Scientific Meeting: College on Problems of Drug Dependence 1995; 57:127.
85. Wallace DR, Mactutus CF, Booze RM. Effects of chronic intravenous cocaine administration on locomotor activity and dopamine D_2/D_3 receptors in rat brain. 57th Annual Scientific Meeting: College on Problems of Drug Dependence 1995; 57:149.
86. Costall B, Lim SK, Naylor RJ. Characterisation of the mechanisms by which purported dopamine agonists reduce spontaneous locomotor activity of mice. Eur J Pharmacol 1981; 73:175-188.
87. Westfall TC, Besson M-J, Giorguieff M-F et al. The role of presynaptic receptors in the release and synthesis of [^{3}H]dopamine by slices of striatum. Naunyn Schmiedebergs Arch Pharmacol 1976; 292:279-287.
88. Silvia CP, King GR, Lee TH et al. Intranigral administration of D_2 dopamine receptor antisense oligodeoxynucleotides establishes a role for nigrostriatal D_2 autoreceptors in the motor actions of cocaine. Mol Pharmacol 1994; 46:51-57.

CHAPTER 7

ELECTROPHYSIOLOGICAL CORRELATES OF IN VIVO ANTISENSE KNOCKOUT OF DOPAMINE D_2 AUTORECEPTORS ON SUBSTANTIA NIGRA DOPAMINERGIC NEURONS

James M. Tepper, Bao-Cun Sun, Lynn P. Martin and Ian Creese

INTRODUCTION

The midbrain dopaminergic system is of critical importance to normal cognitive and motor functioning. The loss of projections from substantia nigra pars compacta dopaminergic neurons to the neostriatum produces the tremor, rigidity and akinesia that is characteristic of Parkinson's disease. A disturbance in the regulation of dopaminergic transmission in the forebrain is believed to underlie the devastating cognitive and behavioral impairments in schizophrenia and perhaps other psychoses, and the effectiveness of antipsychotic drugs correlates directly with their affinity for dopamine receptors.[8] Many drugs of abuse, particularly stimulants such as cocaine and amphetamine, produce their

---

Antisense Strategies for the Study of Receptor Mechanisms,
edited by Robert B. Raffa and Frank Porreca. © 1996 R.G. Landes Company.

euphoric effects through interaction with telencephalic projections of the dopamine neurons of the substantia nigra and ventral tegmental area.[16] Understanding the mechanisms that control dopamine release and its consequent physiological effects in forebrain terminal fields is therefore of great importance to our understanding of the biological bases of a number of different normal and abnormal behaviors.

AUTORECEPTOR MODULATION OF DOPAMINERGIC NEURONAL ACTIVITY

The most basic mechanism by which dopaminergic transmission is regulated is by the rate and pattern of activity of dopaminergic neurons. Increases in impulse flow along dopamine axons lead to increases in dopamine release in terminal fields. This has been verified a number of times by several different techniques including in vitro release studies (see ref. 36 for review), and more recently, by in vivo microdialysis[40] or voltammetric[14] detection of dopamine in the neostriatum, nucleus accumbens or neocortex. One of the more unusual properties of dopaminergic neurons is that they possess receptors for their own neurotransmitter, dopamine, in the somatodendritic region as well as in their axon terminal regions. These receptors are termed "autoreceptors" and serve to modulate dopaminergic neurotransmission through at least two different mechanisms. The somatodendritic autoreceptors are stimulated by dopamine released from presynaptic dendrites of dopaminergic neurons[6,11,30] at dendrodendritic synapses and perhaps at other sites as well.[15,51] These autoreceptors are believed to play a role in self-inhibition of dopamine neurons[17] by opening a potassium conductance that hyperpolarizes dopaminergic neurons[22,23] thereby reducing their spontaneous firing rate. The nerve terminal autoreceptors are stimulated by dopamine released from nearby terminals and act to reduce subsequent impulse dependent dopamine release and dopamine synthesis.[36,52] In a similar manner to that of the somatodendritic autoreceptor, terminal autoreceptor activation produces a hyperpolarization and decrease in excitability of the nerve terminal.[44,45]

It has been known since the late 1970s that there are two major subtypes of dopamine receptors, termed D_1 and D_2 receptors. These subtypes were originally defined on the basis of their differential affinities for various ligands and linkage to intracellular second messenger pathways.[20] The D_1 subtype is linked to stimulation of adenylate cyclase, whereas the D_2 receptor is negatively or not coupled to this enzyme. The binding sites of these two classes of receptors are sufficiently different that specific agonists and antagonists exist, making it possible to demonstrate with electrophysiological or biochemical techniques which of these receptor subtypes mediates a given physiological response. On the basis of these criteria, the dopamine somatodendritic autoreceptor was originally identified as a dopamine D_2 receptor on pharmacological,[29] electrophysiological[23] and molecular biological[27] grounds. Similarly, the autoreceptor on the axon terminals of dopaminergic neurons was identified as a D_2 receptor on pharmacological[4,36] and electrophysiological[44,45] grounds.

MODERN VIEW OF DOPAMINE RECEPTOR CLASSIFICATION

More recently, advances in molecular biology have indicated that there are at least five different genes coding for six different subtypes of dopamine receptor.[33] Two of these receptor subtypes (D_{2S}, D_{2L}) consist of isoforms that arise as a result of post-transcriptional modification of a single gene product.[13,28] Thus, rather than two different dopamine receptors, there exist two families of dopamine receptors. The D1 family is comprised of the D_1 and D_5 receptors and the D_2 family of the D_{2L}, D_{2S}, D_3 and D_4 receptors. Although these different dopamine receptor subtypes are differentially distributed throughout the central nervous system, in some areas mRNAs for two or three of them are found within the same region, and sometimes even within the same neuron.[3,7,32,39] This new knowledge forces a re-examination of the "identification" of specific subtypes of dopamine receptors with physiological effects, particularly with respect to the dopamine autoreceptor.

Although there is certainly a receptor of the D_2 family on the cell bodies and at the nerve terminals of dopaminergic neurons, both D_2 and D_3 mRNA have been identified in substantia nigra and ventral tegmental area [3,33] where they are localized to dopaminergic neurons.[35] Dopamine and most "D_2-selective" agonists and antagonists bind to both the D_2 and the D_3 receptor with relatively high affinity.[33] For instance, studies comparing the distribution of D_2 and D_3 receptor mRNA with [^{125}I]iodosulpiride binding reveal that sulpiride labels both D_2 and D_3 receptors.[3] Thus, the identification of the nigral autoreceptor as a D_2 (and exclusively a D_2) receptor is no longer certain.

Although many "selective" dopamine agonists and antagonists have been synthesized and/or identified in recent years, these drugs are effectively selective only between receptors of the D_1 and D_2 families. Although some ligands (e.g., quinpirole, 7-OH DPAT) show differential binding to D_2 and D_3 receptors under optimal in vitro conditions, the Kds for most agonists and antagonists under physiological conditions are within one order of magnitude between D_2 and D_3 receptors.[12,35] Because of this, they cannot be used to discriminate, at least in a physiologically useful manner, among receptor subtypes that share the same or nearly the same binding sites, i.e., members within the D1 or D2 families. The greatest sequence homology among D_{2S}, D_{2L}, D_3 and D_4 receptors and between D_1 and D_5 receptors lies within the transmembrane spanning regions of the proteins, the region that is involved with ligand binding.[33] Thus, although sulpiride serves as a "selective" antagonist at D_2 receptors and SCH23390 as a "selective" antagonist at D_1 receptors, sulpiride cannot be used to discriminate between physiological effects mediated by D_2 or D_3 receptors,[35] and

SCH23390 similarly is not useful for determining if an effect is mediated by a D_1 or a D_5 receptor.

IN VIVO ANTISENSE OLIOGODEOXYNUCLEOTIDE KNOCKOUT

Antisense knockout refers to the ability of specifically designed short sequences of oligodeoxynucleotides (single-stranded DNA) to bind to their complementary mRNA and stop translation, thereby preventing the expression of the protein that the mRNA coded for. Recent studies have demonstrated the feasibility of using in vivo administration of antisense oligodeoxynucleotides to produce knockouts of specific receptor subtypes and/or subunits including muscarinic m_2, $GABA_B$, $NMDA_{R1}$, neuropeptide Y-Y1 and dopamine D_2 receptors.[19,47,48,50,54,55] Antisense oligodeoxynucleotides enter cells both in vitro or in vivo by receptor-mediated endocytosis or nonselective pinocytosis,[25] and bind specifically to their target mRNA and stop protein translation.[53] As the protein is degraded during the course of normal cellular activity, it is not replaced, resulting in a lack of that protein in the cells which had taken up the antisense oligodeoxynucleotide. The double-stranded DNA/RNA hybrid is also a substrate for ribonuclease H-mediated degradation.[49] Regardless of the precise molecular mechanism of action, the antisense knockout technique offers, for the first time, the ability to reduce or eliminate a particular receptor subtype with absolute specificity. Furthermore, because of the limited spread of antisense oligodeoxynucleotides in the brain, it is possible, for the first time, to selectively knock out either pre- or postsynaptic receptors. This technique has recently been successfully applied to the dopamine D_2 receptor to study the behavioral[50,54,55] and electrophysiological effects[26,37,38] of dopamine D_2 receptor knockout. In this chapter we will summarize the electrophysiological consequences of the knockout of dopamine D_2 autoreceptors by in vivo administration of an antisense oligodeoxynucleotide directed against the dopamine D_2 mRNA.

METHODS

ANTISENSE TREATMENT

The antisense oligodeoxynucleotide and random oligodeoxynucleotide control sequences, and the methods for chronic intranigral administration have already been described.[26,37,38,54] In brief, male Sprague-Dawley rats weighing between 150 g and 250 g were anesthetized with a mixture of ketamine (80 mg/kg) and xylazine (15 mg/kg) i.p. and placed in a stereotaxic apparatus. The scalp was reflected and a small burr hole drilled in the skull overlying and lateral to the left substantia nigra. A 28 g stainless steel infusion guide cannula was lowered at a 20° angle and affixed in place with cyanoacrylate glue and dental cement. Following a 24 hour recovery period, a 33 g injection cannula, 1 mm longer than the guide, was filled with saline vehicle, D_2 random oligodeoxynucleotide control sequence or D_2 antisense oligodeoxynucleotide, inserted into the guide cannula and lowered to a position just above the substantia nigra pars compacta. The cannula was joined to a length of teflon tubing connected through a fluid swivel to a microsyringe pump and saline, D_2 random antisense, or D_2 antisense oligodeoxynucleotide (10-20 µg/µl) was infused continuously at 0.1 µl/hour for 6 days while the animals were housed in individual circular Plexiglas cages with ad libitum access to food and water.

The D_2 antisense oligodeoxynucleotide was a 19-mer complementary to codons 2-8 of the D_2 receptor mRNA with sequence 5'-AGGACAGGTTCAGTGGATC-3'[54] The D_2 random oligodeoxynucleotide control consisted of the same bases as in the D_2 antisense in pseudo-random order with 11 of the 19 bases mismatched from the sense mRNA: 5'-AGAACGGCACTTATGGGTG-3'. Both oligodeoxynucleotides consisted of modified "S-oligodeoxynucleotides" in which the phosphodiester backbone of the nucleotide was modified by the inclusion of a phosphorothioate to increase the resistance of the nucleotide to degradation by endogenous nucleases, but which does not prevent uptake into cells.[1] The oligodeoxynucleotides were synthesized by Oligos Inc., (Wilsonville, OR).

ELECTROPHYSIOLOGICAL MEASUREMENTS

On the 7th day after the start of the infusion, rats were anesthetized with urethane (1.3 g/kg, i.p.), the left femoral vein or a lateral tail vein was cannulated, and the rat installed into a stereotaxic frame. A bipolar stimulating electrode was placed in the ipsilateral neostriatum and extracellular recordings of antidromically identified substantia nigra dopaminergic neurons were obtained by conventional means as described previously.[46]

Dopaminergic neurons were identified by their extracellular waveforms, often characterized by a prominent notch in the initial positive phase and having a duration of 2-5 msec, slow spontaneous activity and long latency antidromic responses evoked from neostriatum that consisted mostly of initial segment only spikes.[9,18,44] The firing pattern of each neuron was classified as pacemaker, random or bursty on the basis of the neuron's autocorrelation histogram.[43] The threshold current for each neuron was defined as the minimum stimulating current that evoked antidromic responses from neostriatum to 100% of the stimulus deliveries.[44]

Following the establishment of a stable baseline firing rate for at least 5 minutes, a dose of apomorphine hydrochloride that was double the previous dose was injected intravenously every two minutes, starting with either 1 or 2 µg/kg. This was continued until complete inhibition of spontaneous activity was obtained, a cumulative dose of 2048 µg/kg was reached, or until the cell was lost. In some cases in which complete inhibition was obtained, haloperidol lactate (50 µg/kg, i.v.) was subsequently administered in an attempt to reverse the inhibition.

AUTORADIOGRAPHY AND HISTOLOGY

For D_2 receptor autoradiography, animals were euthanized by overdose of urethane and the brains rapidly removed and frozen in liquid nitrogen. Subsequently, 20 µm coronal sections were taken on a cryostat at -18°C, thaw-mounted on gelatin-subbed slides and stored at -80°C. Sections were gradually brought to room temperature and incubated for 30 minutes in 0.2 nM [^{3}H] spiperone in 50 mM Tris-HCl buffer, containing 120 mM NaCl, 2 mM $CaCl_2$, 5 mM KCl and 1 mM $MgCl_2$. The incubation period was terminated by rinsing the slides twice for five minutes with ice-cold buffer. Nonspecific binding was determined by incubation in the presence of 1 µM (+) flupentixol. After washing the slides were dipped quickly in ice-cold water and dried under a stream of cold air. Slides were then placed in X-ray cassettes together with [^{3}H] microscales (Amersham) and exposed to Hyperfilm-[^{3}H] (Amersham) for a period ranging from 4 days to 6 weeks at 4°C. Average binding densities were determined with a computerized image analysis system (MCID).

In order to determine if there were any nonspecific neurotoxic effects of the antisense treatment, some brains were sectioned at 60 µm and processed for Nissl staining (Neutral Red) in order to compare the cytoarchitecture of the infused side with that of the contralateral control. In other cases, immunostaining for tyrosine hydroxylase, a marker for dopaminergic neurons in the midbrain, was performed to examine dopaminergic neurons specifically by procedures previously described.[41]

RESULTS

SPONTANEOUS ACTIVITY AND ANTIDROMIC RESPONSES

Spontaneous activity was assessed by measuring spontaneous firing rate, mean interspike interval (ISI), coefficient of variation of the ISI (CV), and by constructing autocorrelation histograms from spontaneous spike trains. The pattern of firing of each neuron was categorized as pacemaker, random, or bursty according to the autocorrelation histogram as described previously.[43] Treatment with D_2 antisense did not significantly alter the mean firing rate, coefficient of variation or the firing pattern of dopaminergic neurons as shown in Table 7.1.

In all neurons that were antidromically activated from ipsilateral neostriatum, the threshold current was measured. In contrast to the other parameters measured, antisense treatment produced a significant reduction in threshold current (Table 7.1), signifying an increase in dopamine terminal excitability, a parameter that we have previously shown is inversely related to the degree of terminal autoreceptor activation.[42]

RESPONSE TO ADMINISTRATION OF APOMORPHINE

Administration of the D_2 antisense oligo for six days markedly attenuated the ability of intravenously administered apomorphine to in-

Table 7.1. Effects of D_2 antisense treatment on electrophysiological properties of nigrostriatal dopaminergic neurons

	Untreated	Saline	Contralateral	D_2 Random Antisense	D_2 Antisense
Mean Firing Rate	3.85 ± 0.32 (17)	3.09 ± 0.26 (10)	3.77 ± 0.50 (15)	3.29 ± 0.39 (11)	4.10 ± 0.27 (55)
Coefficient of Variation	0.342 ± 0.052 (17)	0.435 ± 0.093 (10)	0.504 ± 0.070 (15)	0.405 ± 0.036 (11)	0.439 ± 0.029 (55)
Firing Pattern (P/R/B)	4/6/7	3/5/2	5/6/4	3/6/2	12/27/16
Threshold Current (mA)	1.69 ± 0.31 (14)	1.19 ± 0.19 (6)	2.09 ± 0.35 (6)	1.58 ± 0.33 (11)	1.06 ± 0.17* (18)

* significantly different from pooled controls, Bonferroni/Dunn, $p < .05$

Effects of D_2 antisense treatment on electrophysiological properties of nigrostriatal dopaminergic neurons. Spontaneous firing rates (spikes/sec), coefficient of variation (standard deviation of the mean interspike interval/ the mean interspike interval) and threshold current are expressed as mean ± standard error. Numbers in parentheses refer to number of neurons per group. Firing patterns are expressed as the number of cells firing in (P)acemaker mode, (R)andom mode, and (B)ursty mode respectively for each treatment group and were analyzed by χ^2 analysis.

hibit the firing of dopaminergic neurons as shown for three representative neurons in Figure 7.1. In neurons recorded from untreated controls (A) or ipsilateral to random D_2 oligodeoxynucleotide infusion (B), sequential intravenous injections of apomorphine produced marked inhibition of firing of antidromically identified nigrostriatal dopaminergic neurons before eventually producing complete suppression of spontaneous activity, typically by the time the dose reached 8 µg/kg. This inhibition could be readily reversed by subsequent administration of 50 µg/kg haloperidol, i.v. In contrast, neurons ipsilateral to infusion of D_2 antisense oligodeoxynucleotide were far less susceptible to the inhibitory effects of apomorphine, and in some cases could not be completely inhibited even at a cumulative dose over 2,000 µg/kg. The dose response curves in Figure 7.2 reveal that D_2 antisense treatment produced a dramatic shift to the right in the apomorphine dose-response relation, whereas neurons from rats treated with saline, the D_2 random oligodeoxynucleotide or recorded contralateral to D_2 antisense infusion were indistinguishable from neurons recorded from untreated control rats.

It is worth noting that there was some variability in the response of individual neurons from different antisense-treated animals. About half of the neurons showed a maximum inhibition of firing to about

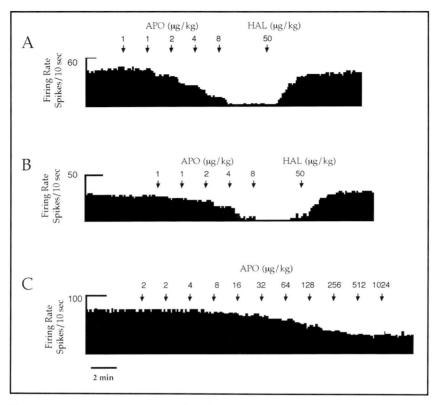

Fig. 7.1. Representative ratemeter plots showing the effects of intravenous administration of the dopamine receptor agonist, apomorphine, on the extracellularly recorded firing rate of antidromically identified nigrostriatal dopaminergic neurons in vivo from saline treated, random antisense treated and D_2 antisense treated rats. (A) Recording obtained from a neuron ipsilateral to saline control infusion for six days. The neuron is inhibited by the bolus of 8 µg/kg. The inhibition is completely reversed by 50 µg/kg of the dopamine D2 class antagonist, haloperidol. (B) Recording obtained from a neuron ipsilateral to D_2 random antisense control infusion for six days. The neuron is almost completely inhibited after 4 µg/kg is and completely inhibited by the bolus of 8 µg/kg. The inhibition is completely reversed by 50 µg/kg, haloperidol. (C) Recording obtained from a neuron ipsilateral to D_2 antisense infusion for six days. The neuron shows virtually no effect to the bolus of 8 µg/kg and is inhibited by less than 50% after a cumulative dose of 512 µg/kg. Complete inhibition could not be achieved even at a cumulative dose of 2,048 µg/kg. In each cell, apomorphine was injected at a dose doubling the previous dose every two minutes.

80% of the pre-drug control levels at the highest dose of apomorphine tested (a bolus of 1024 µg/kg), whereas other neurons could be inhibited to a greater extent, sometimes completely, albeit always at doses much greater than those required to completely inhibit control neurons.

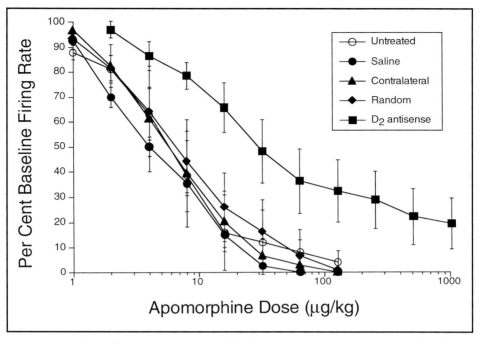

Fig. 7.2. Cumulative dose response curves of the firing rate response to intravenously administered apomorphine in nigrostriatal dopaminergic neurons recorded in untreated control animals, contralateral to D_2 antisense infusion, or ipsilateral to saline infusion, D_2 random antisense infusion or D_2 antisense. In all control groups, neurons exhibited an ED_{50} of approximately 6 µg/kg, whereas the treated neurons' dose response curve is shifted markedly to the right, with an ED_{50} of approximately 31 µg/kg. See text for further details.

AUTORADIOGRAPHY, HISTOLOGY AND IMMUNOCYTOCHEMISTRY

Estimates of the extent of D_2 receptor loss in the pars compacta of the substantia nigra ipsilateral to D_2 antisense infusion obtained from quantitative autoradiography ranged from a low of 40% to a high of approximately 80%, and varied from animal to animal and section to section. The mean changes in binding over the entire pars compacta region for all brains analyzed are shown in Table 7.2. In some cases there appeared to be a near-total absence of D_2 binding near the infusion site in substantia nigra. This was not due to a nonspecific loss of neurons due to neurotoxicity or to nonspecific loss of other receptor binding sites since both Nissl stain and tyrosine hydroxylase immunocytochemistry revealed a normal complement of dopaminergic neurons on the infused side, as shown in Figure 7.3. Furthermore, in other studies, D_1 receptor binding on the infused side was shown to be normal.[37] There were no cases in which the knockout spread to the contralateral noninfused side, and the autoradiograms

Table 7.2. Effects of D_2 antisense treatment on dopamine D_2 receptor binding in substantia nigra pars compacta

	Contralateral	Ipsilateral	Per Cent Change
D_2 Random Antisense	6.53 ± 0.61	6.61 ± 0.72	+ 1.0%
D_2 Antisense	6.58 ± 0.55	3.10 ± 0.49	− 52.9%*

* significantly different from pooled controls, Bonferroni/Dunn, p < .05

Effects of D_2 antisense treatment on dopamine D2 receptor binding in substantia nigra pars compacta. D2 receptors were labeled by 1.2 nM ^{3}H-spiperone with or without 1 µm flupentixol. Numbers under contralateral (control) and ipsilateral (D_2 antisense treated) refer to mean density + standard error in nCi/mg-tissue.

showed a rather sharp delineation of receptor loss near the infusion site in the ipsilateral substantia nigra.

DISCUSSION

These results demonstrate that in vivo supranigral infusion of a phosphorothioated antisense oligodeoxynucleotide directed against the 2nd through 8th codons of the D_2 dopamine receptor mRNA for six days produces a marked reduction in dopamine D_2 receptor binding largely constrained to the substantia nigra with no accompanying signs of neurotoxicity. This reduction in D_2 receptor binding was associated with significant changes in some of the electrophysiological properties of nigrostriatal dopaminergic neurons.

The marked changes in the autoreceptor-mediated properties of dopaminergic neurons after D_2 antisense treatment demonstrates that at least some of the autoreceptors are of the D_2 subtype. Whereas this finding is certainly not unexpected based on previous pharmacological studies of midbrain dopaminergic neurons,[22,23] as mentioned in the introduction, the knowledge that midbrain dopamine neurons also express D_3 receptor mRNA, as well as recent reports that putative D_3-preferring agonists were more potent than D_2-preferring agonists at inhibiting dopaminergic neurons[21,24] left open the possibility that the principal functional autoreceptor on dopamine neurons would prove to be a D_3 receptor. Whereas the present results do not rule out the possibility that there may also be a D_3 autoreceptor on dopaminergic neurons, they confirm that there is certainly a D_2 autoreceptor, at both the somatodendritic and nerve terminal regions.

That there is a D_2 somadendritic autoreceptor is demonstrated by the marked shift to the right in the apomorphine dose-response curve. Previous studies have shown that the inhibition of dopaminergic neurons by low doses of systemically administered apomorphine is due to

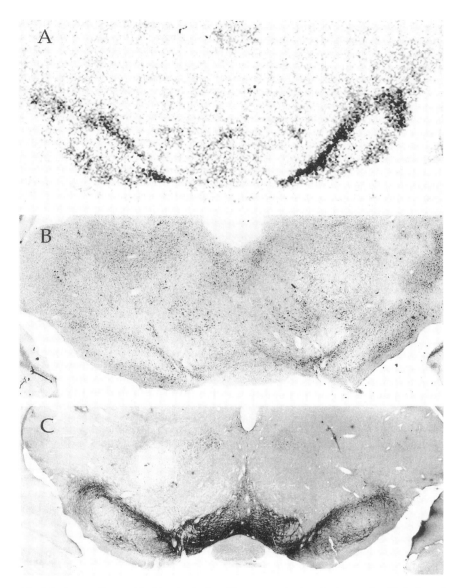

Fig. 7.3. Autoradiographic, histological and immunocytochemical sequela of infusion of D_2 antisense into the left substantia nigra. (A) Specific D2 binding (^{3}H-spiperone which also labels D_3 receptors) is reduced by approximately 70% overall in the substantia nigra (pars compacta plus pars reticulata) of the substantia nigra ipsilateral to the infusion compared to the contralateral side. (B) Nissl staining reveals no apparent damage to neurons in the vicinity of the substantia nigra on the infused (left) side of the brain. (C) Tyrosine hydroxylase immunostaining of a section near to that shown in B from the same brain reveals no loss of dopaminergic (TH positive) neurons on the infused. The lightly stained circular area over the substantia nigra on the left side is nonspecific mechanical damage resulting from the infusion cannula.

a local action at somadendritic autoreceptors on dopaminergic neurons.[2] Thus, the shift in the dose response by the selective reduction in D_2 receptors indicates that there are functional D_2 autoreceptors at the somatodendritic region of nigrostriatal neurons. The inhibition of firing to large doses of apomorphine (> 32 μg/kg) that is still present may represent the action of apomorphine at residual somatodendritic D_2 autoreceptors that still exist after the antisense treatment, since in most cases quantitative autoradiographic analysis of D_2 binding in substantia nigra showed a large decrease, but not a complete elimination of D_2 receptors (see below). However, it is also possible that the residual effects of apomorphine were due to a long-loop postsynaptic effect mediated through forebrain basal ganglia structures, since such output pathways have been shown to be activated by high, but not low doses of apomorphine,[34] and at least part of the inhibitory effects on dopaminergic cell firing of high dose amphetamine, another substance that acts at dopaminergic autoreceptors, has been shown to be due to mediated through these long-loop pathways.[5,31] It seems most likely that both of these explanations contribute to the effects of apomorphine following D_2 antisense treatment. It is also possible that some of the inhibition arises from the activation of D_3 somadendritic autoreceptors, since these would not have been affected by the D_2 antisense treatment, but would be effectively stimulation by apomorphine.

The fact that there was no change in the rate or pattern of spontaneous activity after knockout of D_2 somatodendritic autoreceptors is not consistent with the widely-held belief that these autoreceptors play a role in the self-inhibition of the firing rate of dopaminergic neurons.[17] It is possible that some compensatory changes in the dopaminergic neurons or their afferents occurred that mask the increase in spontaneous firing rate that would be predicted on the basis of the self-inhibition hypothesis. However, this explanation seems unlikely due to the rather short times involved, and because other related properties, including the response to apomorphine and the basal terminal excitability were markedly altered. In addition we have also reported that this same antisense treatment causes an increase in somatodendritic excitability of dopaminergic neurons,[37] a parameter that is also modulated by somatodendritic autoreceptors. Taken together, these data support our previous suggestion, derived from an independent line of evidence, that the activation of somatodendritic autoreceptors on dopaminergic neurons by endogenously released dopamine in vivo may have more to do with local regulation of dendritic excitability than with the direct modulation of spontaneous firing rate.[46]

The decrease in antidromic threshold current in D_2 antisense-treated neurons provides evidence that autoreceptor known to exist on the dopaminergic nerve terminals is a D_2 receptor. Previous studies have shown that acute local administration of the dopamine D_2 class receptor antagonists, haloperidol or sulpiride, into the terminal regions of nigrostriatal neurons causes a reduction in antidromic threshold currents[42,44] because they block the inhibitory effects of endogenous dopamine on the dopaminergic nerve terminals. The decrease in threshold found after antisense treatment in the present study presumably results from a similar loss of D_2 receptor stimulation that arises from the ongoing release of endogenous dopamine. Although these findings need to be confirmed with further experiments in which autoreceptor-mediated changes in terminal excitability and modulation of evoked release of dopamine are measured in antisense treated animals, the present data suggest that there exists an autoreceptor of the D_2 subtype on nigrostriatal terminals, and that application of D_2 antisense oligodeoxynucleotide to the cell body of a neuron results in a loss of D_2 receptors at the nerve terminal as well as the somatodendritic region. As with the somatodendritic autoreceptor data, these findings do not rule out the possibility that there may also be a D_3 autoreceptor at terminal regions of nigrostriatal neurons, but they do demonstrate that a D_2 autoreceptor is present.

While the quantitative autoradiography showed that D_2 receptors in substantia nigra were markedly reduced after nigral infusion of the D_2 antisense, they were not eliminated. The simplest explanation for this is that it may not be possible to achieve complete antisense knockout of dopamine receptors (and perhaps any protein) due to compensatory cellular mechanisms. This might account for the failure of the antisense treatment to completely eliminate the inhibitory effects of apomorphine in the electrophysiological experiments. However, as stated above, different autoradiographic sections through substantia nigra from different animals varied in their degree of receptor loss. The average estimate of the degree of receptor loss (~50%) was determined by comparing the entire area of the substantia nigra pars compacta from many animals. Inspection of individual autoradiograms (see Fig. 7.3A for example) often revealed regions on the antisense-treated side in which binding was nearly indistinguishable from background. This could be due to a highly localized effect of the antisense, in which case averaging over the entire pars compacta would tend give an underestimate of the true degree of knockout at the most strongly affected sites nearest the infusion. In addition, it is certain that some of the bound radioligand represents binding to D_3 receptors, since spiperone labels D_3 receptors with almost the same affinity as D_2 receptors. Thus the true maximal extent of D_2 receptor knockout possible with antisense remains to be determined, but it is certain to be greater than our estimate of 50% based on the quantitative autoradiography reported here.

In summary, these data verify the effectiveness of the technique of in vivo local administration of antisense oligodeoxynucleotides to markedly reduce or eliminate dopamine D_2 receptors in specific cell groups in the brain. Another approach to the problem of identifying the functions of specific dopamine receptor subtypes is the use of transgenic animals in which genes coding for certain receptors have been knocked out.[10] However, the antisense knockout technique may have significant advantages over the use of transgenic knockouts for identifying and studying receptor subtype function when subtype specific agonists or antagonists do not exist. Although the question of the maximal extent of receptor knockout possible with in vivo antisense administration still remains to be determined, because antisense knockout can be directed towards specific areas in the brain, and because the antisense can be administered at any developmental stage or after the brain is fully mature, antisense knockout of CNS receptors avoids the compensatory changes that are likely to occur when a particular receptor is missing from the entire brain from the moment of conception, as is the case with current transgenic models.

ACKNOWLEDGMENTS

This research was supported by HM-52383, MH-52450, a Johnson & Johnson Discovery Award and a Hoechst-Celanese Innovative Research Award.

REFERENCES

1. Agrawal S, Temsamani J, Tang JY. Pharmacokinetics, biodistribution, and stability of oligodeoxynucleotide phosphorothioates in mice. Proc Natl Acad Sci (USA) 1991; 88:7595-7599.
2. Akaoka,H, Charléty P, Saunier C-F, Buda M,& Chouvet G. Inhibition of nigral dopaminergic neurons by sytemic and local apomorphine: possible contribution of dendritic autoreceptors. Neuroscience 1992; 49:879-892.
3. Bouthenet M-L, Souil E, Martres M-P, Sokoloff P, Giros B, Schwartz J-C. Localization of dopamine D_3 receptor mRNA in the rat using in situ hybridization histochemistry: comparison with dopamine D_2 receptor mRNA. Brain Res 1991; 564:203-219.
4. Boyar WC, Altar .A Modulation of in vivo dopamine release by D_2 but not D1 receptor agonists and antagonists. J Neurochem 1987; 48:824-831.
5. Bunney BS, Aghajanian GK. d-Amphetamine-induced depression of central dopamine neurons: Evidence for mediation by both autoreceptors and a strio-nigral feedback pathway. Naunyn-Schmiedeberg's Arch Pharmacol 1978; 304:255-261.
6. Cheramy A, Leviel V, Glowinski J. Dendritic release of dopamine in the substantia nigra. Nature (London) 1981; 289:537-542.
7. Civelli O, Bunzow JR, Grandy DK, Zhou Q-Y, Van Tol HHM. Molecular biology of the dopamine receptors. Eur J Pharmacol (Molecular Pharmacology Section) 1991; 207:277-286.
8. Creese I, Burt DR, Snyder SH. Dopamine receptor binding predicts clinical and pharmacological potencies of antischizophrenic drugs. Science 1976; 192:481-483.

9. Deniau JM, Hammond C, Riszk A, Feger J. Electrophysiological properties of identified output neurons of the rat substantia nigra (pars compacta and pars reticulata): evidence for the existence of branched neurons. Exp Brain Res 1978; 32:409-422.
10. Drago J, Gerfen CR, Lachowicz JE, Steiner H, Hollon TR, Love PE, Jose PA, Ooi GT, Grinberg A, Lee EJ, Huang SP, Bartlett PF, Sibley DR, Westphal H. Altered striatal function in a mutant mouse lacking D_{1A} dopamine receptors. Proc Natl Acad Sci (USA) 1995; 91: 12564-12568.
11. Geffen LB, Jessel TM, Cuello AC, Iversen LL. Release of dopamine from dendrites in rat substantia nigra. Nature (London) 1976; 260: 258-260.
12. Gingrich JA, Caron MG. Recent advances in the molecular biology of dopamine receptors. Ann Rev Neurosci 1993; 16:299-321.
13. Giros B, Sokoloff P, Martres M-P, Riou J-F, Emorine LJ, Schwartz JC. Alternative splicing directs the expression of two D_2 dopamine receptor isoforms. Nature (London) 1989; 342:923-926.
14. Gonon F G, Buda M J. Regulation of dopamine release by impulse flow and by autoreceptors as studied by in vivo voltammetry in the rat striatum. Neuroscience 1985; 14:765-774.
15. Groves PM, Linder JC. Dendro-dendritic synapses in substantia nigra: descriptions based in analysis of serial sections. Exp Brain Res 1983; 49:209-217.
16. Groves PM, Tepper JM. Neuronal mechanisms of action of amphetamine. In: Creese I, ed. Stimulants: Neurochemical Behavioral and Clinical Perspectives. New York; Raven Press, 1983:81-119.
17. Groves PM, Wilson CJ, Young SJ, Rebec GV. Self-inhibition by dopaminergic neurons. Science 1975; 190:522-529.
18. Guyenet PG, Aghajanian GK. Antidromic identification of dopaminergic and other output neurons of the rat substantia nigra. Brain Res 1978; 150:69-84.
19. Holopainen I, Wojcik WJ. A specific antisense oligodeoxynucleotide to mRNAs encoding receptors with seven transmembrane spanning regions deceases muscarinic m2 and γ-Aminobutyric acidB receptors in rat cerebellar granule cells. J Pharmacol Exp Therapeut 1995; 264:423-430.
20. Kebabian JW. Calne DB. Multiple receptors for dopamine. Nature (London) 1979; 277:93-96.
21. Kreiss DS, Bergstron DA, Gonzalez AM, Huang K-X, Sibley DR, Walters JR. Dopamine receptor agonist potencies for inhibition of cell firing correlate with dopamine D_3 receptor binding affinities. Eur J Pharmacol 1995; 277:209-214.
22. Lacey MG. Neurotransmitter receptors and ionic conductances regulating the activity of neurones in substantia nigra pars compacta and ventral tegmental area. In: Arbuthnott GW, Emson PC eds. Chemical Signalling in the Basal Ganglia, Progress in Brain Research. Volume 99. 1993:251-276.

23. Lacey MG, Mercuri NB, North RA. Dopamine acts on D_2 receptors to increase potassium conductance in neurones of the rat substantia nigra zona compacta. J Physiol (London) 1987; 392:397-416.
24. Lejeune F, Millan MJ. Activation of dopamine D_3 autoreceptors inhibits firing of ventral tegmental dopaminergic neurones in vivo. Eur J Phamacol 195; 275:R7-R9.
25. Loke SL, Stein CA, Zhang XH, Mori K, Nakanishi M, Subasinghe C, Cohen JS, Neckers LM. Characterization of oligonucleotide transport into living cells. Proc Natl Acad Sci (USA) 1989; 86: 3474-3478.
26. Martin LP, Kita H, Sun B-C, Zhang M, Creese I, Tepper JM. Electrophysiological consequences of D_2 receptor antisense knockout in nigrostriatal neurons. Soc Neurosci Abst 1994; 20:908.
27. Meador-Woodruff JH, Mansour A, Healy DJ, Kuehn R, Zhou Q-Y, Bunzow JR, Akil H, Civelli O, Watson SJ. Comparison of the distributions of D_1 and D_2 dopamine receptor mRNAs in rat brain. Neuropsychopharmacology 1991; 5:231-242.
28. Monsma Jr FJ, McVittie LD, Gerfen CR, Mahan LC, Sibley DR. Multiple D_2 dopamine receptors produced by alternative RNA splicing. Nature (London) 1989; 342:926-929.
29. Morelli M, Mennini T, Di Chiara G. Nigral dopamine autoreceptors are exclusively of the D_2 type: quantitative autoradiography of [125I]iodosulpiride and [125I]SCH23892 in adjacent brain sections. Neuroscience 1988; 27:865-870.
30. Paden C, Wilson CJ, Groves PM. Amphetamine-induced release of dopamine from the substantia nigra in vitro. Life Sci 1976; 19: 1499-1506.
31. Sasaki K, Suda H, Watanabe H, Yagi H. Involvement of the entopeduncular nucleus and the habenula in methamphetamine-induced inhibition of dopamine neurons in the substantia nigra of rats. Brain Res Bull 1990; 25:121-127.
32. Schwartz JC, Giros B, Martres MP, Sokoloff P. The dopamine receptor family: molecular biology and pharmacology. Seminars in The Neurosciences 1992; 4:99-108.
33. Sibley DR, Monsma Jr FJ. Molecular biology of dopamine receptors. Trends Pharmacol Sci 1992; 13:61-69.
34. Skirboll LR, Grace AA, Bunney BS. Dopamine auto- and postsynaptic receptors: electrophysiological evidence for differential sensitivity to dopamine agonists. Science 1979; 206:80-82.
35. Sokoloff P, Giros B, Martres M-P, Bouthenet M-L, Schwartz J-C. Molecular cloning and characterization of a novel dopamine receptor (D_3) as a target for neuroleptics. Nature (London) 1990; 347:146-151.
36. Starke K, Gothert M, Kilbinger H. Modulation of neurotransmitter release by presynaptic autoreceptors. Physiol Rev 1989; 69:864-989.
37. Sun B-C, Creese I, Tepper JM. Electrophysiology of antisense knockout of D_2 and D_3 dopamine receptors in nigrostriatal dopamine neurons. Soc Neurosci Abstr 1995; 21:1661.

38. Sun B-C, Martin LP, Creese I, Tepper JM. Functional roles of dopamine D_2 autoreceptors in nigrostriatal neurons analyzed by antisense knockout in vivo. 1996; Submitted for publication.
39. Surmeier DJ, Eberwine J, Wilson CJ, Cao Y, Stefani A, Kitai ST. Dopamine receptor subtypes colocalize in rat striatonigral neurons in dopaminergi. Acad Sci 1992; 89:10178-10182.
40. Tepper JM, Creese I, Schwartz DS. Stimulus-evoked changes in neostriatal dopamine levels in awake and anesthetized rats as measured by microdialysis. Brain Res 1991; 559:283-292.
41. Tepper JM, Damlama M, Trent F. Postnatal changes in the distribution and morphology of rat substantia nigra dopaminergic neurons. Neuroscience 1994; 60:469-477.
42. Tepper JM, Groves PM, Young SJ. The neuropharmacology of the autoinhibition of monoamine release. Trends Pharmacol Sci 1985; 6:251-256.
43. Tepper JM, Martin LR, Anderson DR. GABAA receptor-mediated inhibition of nigrostriatal dopaminergic neurons by pars reticulata projection neurons. J Neurosci 1995; 5:3092-3103.
44. Tepper JM, Nakamura S, Young SJ, Groves PM. Autoreceptor-mediated changes in dopaminergic terminal excitability: effects of striatal drug infusions. Brain Res 1989; 309:317-333.
45. Tepper JM, Sawyer SF, Nakamura S, Groves PM. Dopaminergic and serotonergic terminal excitability: effects of autoreceptor stimulation and blockade. In: Feigenbaum JJ, Hanani M, eds. The Presynaptic Regulation of Neurotransmitter Release: A Handbook. Tel Aviv; Freund Publishing Co., 1991:523-550.
46. Trent F, Tepper JM. Dorsal raphé stimulation modifies striatal-evoked antidromic invasion of nigral dopaminergic neurons in vivo. Exp Brain Res 1991; 84:620-630.
47. Wahlestedt C, Merlo E, Koob GF, Yee F, Heilig M. Modulation of anxiety and neuropeptide Y-Y1 receptors by antisense oligodeoxynucleotides. Science 1993; 259:528-531.
48. Wahlestedt C, Golanov E, Yamamoto S, Yee F, Ericson H, Yoo H, Inturrisi CE, Reis DJ. Antisense oligodeoxynucleotides to NMDA-R1 receptor channel protect cortical neurons from excitotoxicity and reduce focal ischaemic infarctions. Nature 1993; 363:260-263.
49. Walder RY, Walder JA. (1988) Role of RNase H in hybrid-arrested translation by antisense oligonucleotides. Proc Natl Acad Sci (USA) 1988; 85:5011-5015.
50. Weiss B, Zhang S-P, Zhou L-W. Antisense oligodeoxynucleotide inhibits D_2 dopamine receptor-mediated behavior and D_2 mRNA. Neuroscience 1993; 55:607-612.
51. Wilson CJ, Groves PM, Fifková,E. Monoaminergic synapses, including dendro-dendritic synapses in the rat substantia nigra. Exp Brain Res 1977a; 30:161-174.

52. Wolf ME, Roth RH. Autoreceptor regulation of dopamine synthesis. In: Kalsner S, Westfall TC, eds. Presynaptic Autoreceptors and the Question of the Autoregulation of Neurotransmitter Release. Ann. New York Acad Sci 1990; 604:323-343.
53. Yakubov LA, Deeva EA, Zarytova VF, Ivanova EM, Ryte S, Yurchenk LV, Vlassov, VV. Mechanism of oligonucleotide uptake by cells: involvement of specific receptors? Proc Natl Acad Sci (USA) 1989; 86: 6454-6458.
54. Zhang M, Creese I. Antisense oligodeoxynucleotide reduces brain dopamine D_2 receptors: behavioral correlates. Neurosci Lett 1993; 161:223-226.
55. Zhou L-W, Zhang S-P, Qin, Z-H, Weiss B. In vivo administration of an oligodeoxynucleotide antisense to the D_2 dopamine receptor messenger RNA inhibits D_2 dopamine receptor-mediated behavior and the expression of D_2 dopamine receptors in mouse striatum. J Pharmacol Exp Ther 1994; 268:1015-1023.

CHAPTER 8

c-FOS ANTISENSE: INFUSION AND EFFECTS

Deborah Young and Michael Dragunow

INTRODUCTION

One of the current challenges in molecular neurobiology is to identify the signal transduction cascade that leads to permanent changes in neuronal function. The immediate early genes (IEGs) are thought to play an important part in this cascade as some of them encode inducible transcription factors which are involved in regulating changes in expression of late effector genes which cause phenotypical changes in neurons. However, although much circumstantial evidence exists implicating their role in a variety of different brain functions, there is no actual conclusive evidence that IEGs are involved in any particular function in the brain. The recent development of antisense (AS) oligodeoxynucleotide technology may revolutionize our approach to studying the role of IEGs in the central nervous system and has the potential to be a powerful tool for directly elucidating specific gene function in vivo. By using AS oligonucleotides to knock out a specific IEG and hence production of its protein product in vivo, one could then demonstrate that IEG activation leads to a particular physiological consequence. In this chapter, we will focus on some of the in vivo applications of AS oligonucleotides, in particular that of the prototypical IEG *c-fos*. We will begin with a brief description of IEGs and how *c-fos* may function as a transcription factor in neurons. This will be followed by some information about AS oligonucleotides and we will then focus on the in vivo applications of *c-fos* AS in the central nervous system.

IMMEDIATE EARLY GENES

The IEGs are the first set of genes activated by neuronal stimulation and characteristically can be expressed under conditions of protein synthesis inhibition.[1,2] Many IEGs such as those from the Fos and Jun families participate in a fundamental biological control mechanism, the regulation of gene expression. Thus, they are often thought of as "nuclear third-messengers," linking extracellular signals such as those carried by neurotransmitters to changes in neuronal phenotype by regulating the expression of so-called "late-response genes."[3-6]

One of the most extensively studied immediate early genes is c-fos. Fos cannot act as a transcription factor alone but is dependent on the formation of a dimeric complex with members of the Jun family (Jun D, c-Jun, Jun B)[7-9] through the interaction between two highly conserved structural domains on Fos and Jun proteins called the leucine zipper.[8,10] This dimeric complex called activator protein 1 (AP1) regulates target gene expression by enabling AP1 to bind to the AP1 consensus sequence, which is a common transcription regulatory element within the promoter region of various target genes.[4,5] The specificity and affinity of the dimer for the regulatory element and its efficacy in regulating transcription is determined by the constituents of the dimer and hence by the relative contribution of the monomers present, thus Fos/Jun B dimers have different transactivational activity to Fos/c-Jun dimers.[11-14] A number of genes which could be target genes have been identified and include preproenkephalin,[15] GAP-43,[16] cholecystokinin,[17] amyloid precursor protein,[18] neuropeptide Y,[19] dynorphin,[20] tyrosine hydroxylase,[21,22] nerve growth factor,[23-25] Trk B[26] and many others.

c-fos mRNA and protein are rapidly and transiently induced following a wide variety of external stimuli, including pharmacological agents[27-30] or pathophysiological stimuli such as seizures,[27,31,32] cerebral ischemia,[33-38] and traumatic brain injury.[39-42] In addition, c-fos may also play a role in physiological transcriptional control in the brain, as high basal levels are found in areas that function in the processing of sensory input, such as the visual cortex following photic stimulation.[43]

However, although c-fos is induced in response to a wide variety of stimuli, the identity of the target genes which are regulated by Fos transcription factor action are unknown. There has been considerable research devoted to determining the link between immediate-early gene induction and subsequent steps in the pathways that lead to changes in gene expression and permanent change in brain function. However, the elucidation of this part of the pathway has been slow because there have been no pharmacological agents or other in vivo techniques available that can knock out specific IEGs.

ANTISENSE OLIGONUCLEOTIDES

The principle objective underlying antisense technology is to interfere with gene expression by preventing translation of proteins from mRNA. Antisense (AS) oligonucleotides are short fragments of DNA or RNA,[44] typically ranging from between 15-25 nucleotides that are complementary to a portion of the mRNA of interest. Fifteen bases has been established as being the minimum requirement for specific recognition of a single mRNA species and 20-25 being the maximum, as an increase in additional bases may limit cellular uptake.[45] The mechanism of oligonucleotide uptake appears to be by endocytosis,[46,47] perhaps initiated by a receptor-like recognition mechanism.[48,49] The specific hybridization of the oligonucleotide with its complementary target sequence on the mRNA may result in subsequent inhibition of protein translation by at least two possible mechanisms. The first is degradation of RNA by RNase H, which selectively cleaves the RNA at DNA-RNA heteroduplexes and the second mechanism is the arrest of translation initiation caused by AS hybridization to the 5'-untranslated region or the initiation site on the mRNA.[50-53] Generally, AS oligonucleotides are synthesized to sequences in the vicinity of the initiation codon and a number of successful experiments, which appear later in this chapter have used AS oligonucleotides targeted to sequences flanking the AUG initiation codon.[54,55]

The stability of AS oligonucleotides has been a problem for their in vivo use, as unmodified phosphodiester AS oligonucleotides can be rapidly degraded by nucleases present in blood and cerebrospinal fluid.[56,57] Thus high enough concentrations in neurons required to block translation may not be reached. With site specific injections, tissue damage is to be expected, so it may also be preferable to inject metabolically stable oligonucleotides to prevent their degradation. These problems have largely been overcome with the development of AS oligonucleotides with a modified phosphodiester backbone.[57,58] Instead of phosphodiester linkages, AS oligonucleotides have been synthesized with a phosphorothioate backbone. Phosphorothioate oligonucleotides have one of the nonbridging oxygen atoms replaced in the internucleotide linkage by a sulphur atom which renders them more resistant to cleavage by nucleases[44,50,51,59] Whitesell et al[57] recently characterized the uptake, stability and toxicity of phosphorothioate oligonucleotides and found them to be stable in brain after intraventricular administration to rats. Oligonucleotides were taken up by brain cells, especially astrocytes and showed extensive brain penetration with low toxicity. However, problems of nonspecific action have been encountered with these modified oligonucleotides and some in vivo animal studies have shown nonsequence specific effects.[61] A recent adaptation to this has been the development of AS oligonucleotides only partially modified by phosphorothioate linkages and these are expected to act in a more sequence specific manner.[62,63] Other methods developed to protect AS oligonucleotides from nuclease-mediated degradation and improve cellular uptake include packaging the AS oligonucleotide inside a lipid capsule.[64]

To control for the specificity of the AS oligonucleotides several controls can be used. The first are sense analogs of the presumably active oligonucleotide molecule which control for any effects the phosphorothioate backbone may have. However, in most cases the base composition is greatly altered and may have a different biological effect unlike that of the AS oligonucleotide. Therefore a more stringent control may be a mismatched analog which maintains the same base composition as the antisense molecule. An oligonucleotide with 1-4 mismatches will have a 500-fold reduction in affinity for the target sequence of interest.[50,52] Another control is to use several nonoverlapping antisense oligonucleotides that induce similar inhibition of synthesis of the targeted protein. Other controls are to assay for related protein subtypes in the same region, for example if using a *c-fos* AS, then determine whether changes also occur in related family members such as fos B mRNA or protein levels or perhaps changes in another protein known to also exist in the same region and known to respond to stimulation as does the gene of interest.

For a more in depth review of this area, we direct the reader to other chapters in this book and to a number of excellent reviews on antisense oligonucleotides.[45,46,58,65,66]

THE IN VIVO USE OF *C-FOS* ANTISENSE

A few studies have now been conducted which demonstrate that the infusion of AS oligonucleotides into various brain regions (e.g., striatum) will enter neurons and block target mRNA expression. We will now review a number of studies using *c-fos* antisense techniques to determine the role of Fos in mediating the effects of neuronal stimulation in various brain regions.

THE ROLE OF FOS IN BASAL GANGLIA FUNCTION

The largest number of in vivo AS studies have been conducted in the striatum which is a component of the basal ganglia. The major function of the basal ganglia which are located next to the lateral ventricles in the brain is the control of conscious movement. One of the pathways which plays a crucial role in controlling motor functions is the dopaminergic projection from the substantia nigra to the neostriatum (the nigrostriatal pathway). Loss of the dopaminergic nigrostriatal projections, as demonstrated by histological studies, has been postulated to result in the symptoms exhibited by sufferers of the chronic neurodegenerative disorder Parkinson's disease, such as impaired coordination (ataxia) and difficulty in initiating movements (dyskinesia). Thus unilateral lesioning of the nigrostriatal dopaminergic pathway by 6-hydroxydopamine (6-OHDA) is a commonly used animal model of Parkinson's disease.[67,68] Lesioned animals exhibit normal locomotor behavior and no asymmetries, however stimulation of motor activity by activation of the dopaminergic nigrostriatal neurons by indirect (e.g., amphetamine, cocaine) or direct (e.g., apomorphine) acting D_1 dopamine agonists produces rotational behavior.[68-75] Immunohistochemical studies of brain sections taken from these animals have demonstrated induction of Fos protein in striatal neurons only on the intact side for amphetamine,[70,74] and only on the lesioned side for apomorphine,[70,73,76] the sides responsible for stimulation of rotation. Thus these results have implicated Fos expression in the mediation of the locomotor stimulant effects of dopamine agonists.

In subsequent studies, various research groups have used c-fos AS techniques in an attempt to provide further support for this hypothesis. The first in vivo demonstrations of suppression of Fos production by c-fos AS was conducted in the striatum.[54] In this pioneering study, c-fos AS phosphorothioate oligonucleotides targeted to a sequence that flanked the initiation codon (5'-129-143-3')[77] of the c-fos gene with the sequence 5'GAA-CAT-CAT-GGT-CGT-3' were infused unilaterally into the striatum 10 h prior to treatment with the indirect-acting D_1 dopamine receptor agonist amphetamine, which previous studies had shown rapidly and transiently induced Fos in the striatum when injected alone.[29,70] Infusion of the AS selectively suppressed amphetamine-induced Fos production in an area encompassing 20-50% of the striatum, while Fos production in the opposing striata was not suppressed by control c-fos sense infusions or in uncannulated amphetamine-treated animals.[54] The specificity of the AS was confirmed by the observation that the expression of another IEG, NGFIA (also known as zif268, Krox-24, egr-1) was not reduced in striatal neurons on the AS side. Subsequent studies have shown that both c-fos AS oligonucleotides with an unmodified phosphodiester backbone or partially modified by phosphorothioate linkages are ineffective or have little effect in the same experimental paradigm.[79] Furthermore, a different AS directed to the initiation codon and a portion of the coding region with sequence (5'ACG-TAA-CCG-TAA-CCA-3') also had no effect.[79]

Following the initial study by Chiasson et al,[54] other studies showed that using the same c-fos AS oligonucleotide sequence, infusion into the striatum or nucleus accumbens prevented Fos protein expression and blocked the locomotor stimulant effects of indirect acting D_1-dopamine agonists (i.e., cocaine, amphetamine) in rats, a similar effect to that observed in amphetamine-challenged lesioned animals.[55,79] Sense c-fos oligonucleotide infusions did not have any effects on Fos expression.[79] Infusion into the striatum of c-fos AS and c-fos sense into the opposing striata, 10 h prior to the administration of amphetamine induced rotational behavior towards the AS injected side of the brain and bilateral infusion of c-fos AS into the nucleus accumbens blocked the locomotor stimulant effects of cocaine.[55] Rotational behavior only occurred when there was a clear difference in Fos-like immunoreactivity between the two striata.[78] The results from the Sommer[79] and Heilig[55] studies suggested that the induction of c-fos by amphetamine and cocaine was directly involved in their locomotor stimulant effects. However, it was also noted that the locomotor stimulant effects of cocaine occurred at a dose of 10 mg/kg whereas higher doses of 25 mg/kg are required to induce *c-fos* levels in the striatum.[29,55] In addition, the time of onset of rotational behavior occurred 10-15 min after administration of amphetamine or cocaine, which would be too soon for transcription and translation of c-fos mRNA levels. This suggested that the amphetamine-mediated induction of c-fos mRNA and locomotor behavior may be independent of each other. Therefore, perhaps *c-fos* AS infusion into the striatum alters its biochemistry such that the locomotor stimulant effects of cocaine and amphetamine, in addition to the effects on *c-fos* induction, are prevented.[80] This hypothesis was supported by the finding that the direct-acting dopamine agonist apomorphine which by itself does not induce *c-fos* in normal striatum,[29] also induced rotational behavior towards the AS side.[80] In addition, the expression of JunB which is also induced in striatal neurons when cocaine and amphetamine are administered alone, was also suppressed after *c-fos* AS infusion even though the same *c-fos* AS sequence used did not suppress amphetamine-induced NGFIA expression.[54] Thus it seems unlikely that *c-fos* forms part of the dopamine-mediated signal transduction pathway that generates locomotion[80] although activation of D_1-dopamine receptors are involved in both locomotor behavior and *c-fos* induction. Another possibility is that *c-fos* AS may be acting to switch off genes that code for proteins that normally mediate amphetamine induction of both *c-fos* and JunB in striatal neurons (e.g., the D_1-dopamine receptor).[80] However, a further investigation involving determination of the effect of *c-fos* AS on various neurochemicals involved in aspects of striatal function following an apomorphine challenge showed that there were no changes in D_1 or D_2 dopamine receptors, $A2_a$ adenosine receptors, cannabinoid receptors or tyrosine hydroxylase, somatostatin, enkephalin, dynorphin or substance P, although basal levels of Krox 24 were reduced in striatal neurons.[81] The significance of this effect is unknown at present. Hooper et al[78] have used the same *c-fos* AS to show little or no change in striatal egr-1 expression following *c-fos* AS after amphetamine administration. The reasons for these different results is unclear. A recent study has demonstrated that striatal *c-fos* induction by amphetamine is regulated by phosphorylation of the constitutively expressed transcription factor cyclic-AMP-response element binding protein (CREB),[82] thus it is possible that the *c-fos* AS may be having an effect on CREB.

While psychostimulants such as amphetamine and cocaine increase *c-fos* production by activation of D_1 dopamine receptors in the striatum, *c-fos* induction by antipsychotic agents such as haloperidol are thought to be mediated by D_2 dopamine receptor mechanisms.[83-87] The endogenous neuropeptide neurotensin is thought to function as a neuromodulator or neurotransmitter and is widely distributed throughout the central nervous system.[88-91] Exogenously applied neurotensin to animals results in behavioral, biochemical and electrophysiological effects resembling that of haloperidol and it has been postulated that neurotensin is an endogenous neuroleptic-like compound.[92] The acute or chronic administration of haloperidol causes an increase in neurotensin/neuromedin N mRNA and protein in the dorsolateral striatum and nucleus accumbens.[93–102] However, the mechanisms by which haloperidol increases striatal neurotensin/neuromedin N gene expression is unknown. Because studies have identified several regulatory elements in the 5' region of the neurotensin/neuromedin N gene including AP1, glucocorticoid response element (GRE) and cAMP response elements (CRE),[103] it has been suggested that Fos and Jun proteins may be involved in the haloperidol-induced activation of this gene. Further support for this supposition includes the colocalisation of Fos immunoreactivity in a large proportion of neurons with neurotensin-like immunoreactivity following induction by haloperidol[86,100,101,104,105] and the temporal profile of induction of Fos precedes an increase in neurotensin mRNA levels.[100,101,106,107]

Using c-fos AS techniques two studies have been conducted to determine whether increases of neurotensin/neuromedin N mRNA levels after haloperidol are mediated by Fos. In the first, Merchant et al[108] infused a phosphorothioate c-fos AS or sense with the same sequence as that used by Chiasson et al[54] into the central core of the neostriatum. c-fos AS injected 8h prior to a challenge with haloperidol resulted in a 50% attenuation of the increase in neurotensin/neuromedin N mRNA in the dorsolateral striatum on the same side, while sense injections in the contralateral side did not affect the increase in neurotensin/neuromedin N mRNA in this region. A similar effect of c-fos AS was also found by Robertson et al[109] following intrastriatal infusion of two c-fos AS oligonucleotides different in sequence to that used by Chiasson et al[54] These two c-fos AS oligonucleotides were targeted to sequences complementary to bases 109-126 (5'-TTT-GGG-CAA-AGC-TCG-GCG-3', called AS1) and 127-144 (5'-AGA-ACA-TCA-TGG-TCG-TGG-3', called AS2) of the c-fos gene.[77] Haloperidol administered to control animals induced Fos, Jun B, Fos B and NGFIA-like immunoreactivity bilaterally in striatal neurons. While Fos-like immunoreactivity was suppressed in striatal neurons in animals receiving c-fos AS infusions 14 h prior to the haloperidol challenge by 65% for the AS1 oligonucleotide and 45% for the AS2 oligonucleotide, the expression of Fos B, Jun B, Jun and NGFIA was not suppressed on the side ipsilateral to AS, nor was the expression of these IEGs and Fos suppressed on the contralateral sense infusion side. All three AS oligonucleotides used in these two studies also reduced haloperidol-induced neurotensin/neuromedin N mRNA levels, while proenkephalin mRNA which is also expressed in these striatal neurons was unaffected by the sense or AS infusions. That proenkephalin gene expression was unaltered by c-fos AS is consistent with reports that cAMP-response element binding protein (CREB) but not Fos may regulate constitutively and haloperidol-stimulated proenkephalin mRNA levels in striatal neurons.[110,111] These results have provided additional evidence suggesting that neurotensin/neuromedin N gene expression may be regulated by Fos following acute haloperidol administration.

THE ROLE OF FOS IN CONTROL OF BLOOD PRESSURE

The baroreceptor sympathetic reflex is involved in the central control of blood pressure. In response to an elevation in arterial blood pressure, excitatory neurons in the aortic arch and carotid sinus are activated. These neurons project to form synapses with neurons in the nucleus tractus solitarius in the rostral medulla which in turn activate inhibitory neurons (which are thought to use GABA as their neurotransmitter) in the caudal ventrolateral medulla. Depolarization of these inhibitory neurons causes a reduction in firing rate of bulbospinal sympathoexcitatory neurons in the rostral ventral medulla, which project to sympathetic neurons in the spinal cord resulting in a reduction in arterial pressure.[112]

c-fos induction has been implicated in mediating the central control of blood pressure. Following a reduction in blood pressure by stimulation of the aortic depressor nerve (i.e., neuronal pathway that projects from the aortic arch to the nucleus tractus solitarius)[113] or by hemorrhage,[114-116] Fos is induced in numerous brain areas including nuclei in medullary regions such as nucleus tractus solitarius, area postrema, rostral and caudal ventrolateral medulla, nucleus ambiguus, and medullary reticular formation. In the pons, neurons in the locus ceruleus, supraoptic nucleus, bed nucleus of the stria-terminalis, islands of Calleja are some of the regions that show Fos induction.[113]

Chalmers and his group have used *c-fos* AS techniques to examine the role of Fos in mediating blood pressure changes in two of the central pathways that appear to be critical for the control of blood pressure. These are neurons in the rostral ventral medulla which are excitatory[117] and project to neurons in the spinal cord, and the inhibitory neurons in the caudal ventrolateral medulla[118] which innervate the rostral ventral lateral medullary neurons.[119,120] Injection of muscimol, an analog of the inhibitory neurotransmitter GABA into the caudal ventrolateral medulla disinhibits the bulbospinal pressor pathway (i.e., the inhibitory pathway that projects from the caudal ventrolateral medulla to rostral ventral medulla) and increases *c-fos* expression in the rostral ventral medulla.[121,122] A unilateral infusion of a phosphorothioate *c-fos* AS (with same the sequence to that used by Chiasson[54]) into the rostral ventral medulla and sense infusions into the contralateral side, 6 h prior to muscimol administration into the caudal ventrolateral medulla, led to a 50% reduction in Fos immunoreactivity 2 h later in rostral ventrolateral medullary neurons.[122,123] In a second series of experiments, one set of animals received *c-fos* AS bilaterally while another set of animals received sense infusions into the rostral ventral medulla. Basal blood pressure which was monitored hourly gradually fell in rats receiving AS so that at the end of 6h it was 20 mmHg lower than in control sense-injected animals.[122,123] Finally, muscimol was injected into the caudal ventrolateral medulla of these same two groups of animals and blood pressure monitored. Animals receiving c-fos AS had a much attenuated increase in blood pressure. From these results it is postulated that the expression of c-fos in the rostral ventral medulla is involved in the baroreceptor reflex. Similarly, Chiu et al[116] have also shown that the induction of Fos-immunoreactivity in neurons of the nucleus tractus solitarius, ventrolateral medulla, the intermediolateral cell column of the spinal cord and various other regions caused by hemorrhage can also be attenuated by c-fos AS infusion i.c.v. for two days prior to hemorrhage.

THE ROLE OF FOS IN CIRCADIAN RHYTHM AND SLEEP/WAKING CHANGES

The suprachiasmatic nucleus (SCN) located in the hypothalamus is the predominant circadian pacemaker in mammals.[124] The daily light-dark cycle synchronizes the endogenous circadian pacemaker within the SCN to the environmental 24 h rhythm. However, the molecular mechanisms by which light controls the circadian clock are poorly understood.

IEGs such as c-fos may play a role in the control of circadian rhythm. Light-induced phase shifts of circadian rhythm induce c-fos, c-jun, junB and Krox 24 mRNAs in SCN neurons.[125-130] IEGs are expressed in the SCN only during that circadian time when light is capable of shifting the circadian rhythm and IEG expression is mainly restricted to the ventrolateral part of the SCN, the terminal region of retinal afferents.[124]

Using a combination phosphorothioate c-fos/jun-B AS infused i.c.v. into the third ventricle at the level of the ventral SCN, Wollnik et al[132] have investigated whether the light-induced phase shift in circadian rhythm (measured by changes in circadian locomotor activity) are dependent of expression of Fos and Jun-B. The sequence of c-fos AS used was 5'-CGA-GAA-CAT-CAT-GGT-CGA-AG-3' and is the same as that used for inhibiting Fos expression in the spinal cord.[133] A light pulse given to a group of animals at a time previously determined[134] to cause the maximal phase delay of circadian rhythmic activity resulted in a phase shift of the circadian rhythm by approximately -125 min. Animals given a nonsense c-fos/junB control infusion 6 h prior to the light pulse also caused a phase delay similar to that found in animals given the light pulse only. However, c-fos/jun-B AS given 6 h prior to the light-pulse inhibited this phase shift. Animals given the AS only without the subsequent light pulse did not change the endogenous free running circadian rhythm. Fos and Jun B immunoreactivity was induced in SCN neurons in animals that had received the light pulse alone and nonsense infusions. However while the AS infusion specifically blocked Fos and Jun B protein induction in SCN neurons, Fos B and c-Jun immunoreactivity was unaffected. From these results Wollnik et al[132] propose that Fos and Jun B play an important role in the synchronization of environmental cues from the dark-light cycle to the endogenous circadian pacemaker. As yet the target genes for Fos and Jun B that are responsible for the shifted locomotor activity rhythm are unknown.

A closely associated brain region to the suprachiasmatic nucleus, the medial preoptic area (MPA) in the hypothalamus has been implicated to play a central role in the regulation of sleep and waking.[135] c-Fos mRNA and Fos protein are induced in neurons of the MPA after periods of spontaneous wakefulness or sleep deprivation.[136,137] Cirelli et al[138] have investigated the effects of *c-fos* AS infusion (same sequence as that of Chiasson et al[54]) into the zone between medial and lateral preoptic areas on sleep waking patterns in the rat. In animals that received *c-fos* AS, Fos immunoreactivity in nuclei of neurons in the MPA was effectively reduced 11 h after the infusion which was given after several hours of light. Animals receiving *c-fos* AS showed much less sleep than control animals on the day after the injection suggesting that the blockade of Fos protein expression in this brain region may interfere with the mechanisms that assess the duration and intensity of prior wakefulness or with mechanisms that bring about sleep.

THE ROLE OF FOS IN NOCICEPTION

There is much evidence showing induction of IEGs in the spinal cord following noxious stimuli.[139] *c-Fos* is one of the IEGs induced in spinal nociceptive neurons predominantly localized to the superficial laminae (I/II$_o$) but also in deeper lamina following thermal and chemical noxious stimuli.[140-144] For example, injection of formalin into a hind paw of a rat results in a biphasic nociceptive response with accompanying inflammation and *c-fos* induction in the superficial laminae of the lumbar enlargement of the cord ipsilateral to the side of injection.[144] Another consequence of noxious stimulation is the increased expression of opioid peptide gene expression in spinal cord regions involved in nociceptive information processing.[145,146] Opioid peptides such as dynorphin and enkephalin are released following noxious stimuli and electrophysiologically they suppress the activity of dorsal horn neurons[147-149] and in vivo have potent antinociceptive properties.[150] Thus increases in dynorphin in the superficial and deep dorsal horn laminae are involved in modulating nociceptive information from the spinal cord to the brain. Preprodynorphin mRNA and dynorphin-like immunoreactivity are also induced in the same spinal nociceptive neurons that show increased Fos expression.[141,146] As the AP1 binding sequence has been demonstrated in the 5' upstream region of the preprodynorphin gene,[151] Fos may have a role in regulating the expression of this gene and thus its protein product which may then have a role in the progression of the pain response. However, as with other studies, although the expression kinetics and localization of induction suggests that c-fos may be involved in regulating preprodynophin gene expression,[141,146] a causal relationship between c-fos production and an alteration in prepreprodynophin gene expression has yet to be established.

To determine the role in Fos in mediating the pain response, Hunter et al[152] have examined the behavioral effects following intrathecal infusion of phosphorothioate c-fos AS prior to an injection of 5% formalin into the plantar surface of a rat hind paw. A behavioral characteristic following formalin injection is a biphasic licking/biting response of the injured limb consisting of an acute phase of intense activity lasting approximately 20 min followed by a more prolonged but less intense phase lasting about 45 min. c-fos mRNA is induced within 10 min of formalin injection. c-fos AS administered 4 h prior to the formalin challenge inhibited Fos protein and preprodynophin mRNA expression in laminae II/II$_o$ of the dorsal horn ipsilateral to side of injection and significantly increased the licking/biting behavioral response. In comparison, increased Fos expression and no increase in behavioral response was observed in control sense and saline-treated animals. It is interesting to note the specificity of this response as only moderate increases in another opioid peptide, preproenkephalin mRNA also occurred in response to formalin injection, even though the AP1 binding site has also been demonstrated in the promoter of this gene.[15] However, other IEGs such as NGFIA and Jun which are also induced following noxious stimuli[144] may be involved in regulating preproenkephalin gene expression. From these results, these investigators have proposed that Fos expression regulates the late response preprodynophin gene and the subsequent production of this opioid peptide which then mediates the nociceptive response.

Using a partially phosphorothioated c-fos AS (5'-CGA-GAA-CAT-CAT-GGT-CGA-AG-3'), Gillardon et al[133] demonstrated that superfusion of c-fos AS on one side of the spinal cord and c-fos sense control on the contralateral side, 6 h prior to thermal noxious stimuli produced by immersion of both hindpaws into hot water can also significantly attenuate Fos protein expression in dorsal horn neurons on the side that received the AS. Fos protein expression is not affected on the sense infusion side, nor is c-Jun or JunB expression altered on both sides, suggesting the effect is sequence-specific.

THE ROLE OF FOS IN MEDIATING NEURONAL INJURY AND REPAIR

Perhaps one of the most exciting uses for c-fos AS oligonucleotides may be to identify the molecular cascade of events involved in neuronal injury and adaptation of the brain to injury and functional recovery. Fos is rapidly and transiently induced following brain insults such as mechanical injury,[39–42] seizures (status epilepticus),[27,31,153] and focal and global hypoxia-ischemia[33,36,38,154–159] in hippocampal and cortical neurons. A number of potential target genes that have neurotrophic properties and which may be involved in the protection or function in repair mechanisms after neuronal injury include heat shock protein,[154,160,161] amyloid precursor protein,[18] neurotrophins[162] and trk B.[26] In addition, following brief ischemic episodes and status epilepticus, susceptible neuronal populations die by a delayed programmed cell death process termed apoptosis. Apoptosis is characterized by morphological changes such as DNA fragmentation,[163] membrane blebbing, cell shrinkage and may be a protein synthesis-dependent process, as protein synthesis inhibitors have been shown to be neuroprotective.[164–166] Thus it has been hypothesized that there are proteins synthesized after hypoxia-ischemia that may be involved in the subsequent delayed neuronal death process. The synthesis of these "killer proteins" may also be regulated by Fos and Jun proteins. The causal relationship between the Fos/Jun cascade and the subsequent expression of the late response genes is unclear. Thus whether the induction of Fos is "good" or "bad" is unknown.

One study has demonstrated the ability of intraventricular infusion of *c-fos* AS oligonucleotides to penetrate the hippocampus and block focal-ischemia-induced Fos production in hippocampal neurons.[168] An antisense strategy involving infusion of a phosphorothioate *c-fos* AS (5'-CAT-CAT-GGT-CGT-GGT-TTG-GGC-AAA-CC-3') 17-18 h prior to production of focal cerebral ischemia was used. While Fos-like immunoreactivity was induced after focal ischemia in dentate granule and pyramidal neurons of the hippocampus in sham-operated animals and animals receiving sense infusions, Fos production in these neurons was totally suppressed in animals receiving AS. In line with these findings, a post-ischemic increase in AP1-binding activity was also detected in sham-operated animals and animals given a pre-ischemic sense infusion. However, a reduction of AP1 binding activity was detected in animals given pre-ischemic *c-fos* AS. CREB binding activity was unaffected.

While the subsequent effect of Fos inhibition on the survival or death of these neurons was not examined in this study, this pioneering study has been one of the first to demonstrate the in vivo suppression of a specific gene of interest using AS techniques in hippocampal regions, in this case the finding that *c-fos* AS can block Fos production in hippocampal neurons after intraventricular infusion.

FUTURE DIRECTIONS

From the studies conducted thus far, it can be seen that the use of AS technology in vivo is still in its infancy. Nevertheless, in some areas the use of AS technology has advanced our knowledge into the functional role of IEGs in the central nervous system, for example in the basal ganglia. There still appears to be a few problems associated with the use of AS, such as determining the exact criteria essential for optimal suppression of the mRNA of interest. Criteria for optimal suppression of mRNAs in brain regions with laminated structures such as hippocampus are largely unknown, as all areas thus far examined tend to be homogenous brain regions where AS oligonucleotides tend to diffuse quite readily. However, the technique has provided an insight into the functions of IEGs. The AS technique is particularly useful for studying transcription factors such as Fos, Jun and Krox 24, because these are nuclear proteins that cannot be targeted in vivo by other methods such as neutralizing antibodies.

REFERENCES

1. Cochran BH, Zullo J, Verma IM et al. Expression of the c-fos gene and of a fos-related gene is stimulated by platelet-derived growth factor. Science 1984; 226:1080-1082.
2. Greenberg ME, Ziff EG, Greene LA. Stimulation of neuronal acetylcholine receptors induces rapid gene transcription. Science 1986; 234:80-83.
3. Dragunow M, Currie RW, Faull RLM et al. Immediate-early genes, kindling and long-term potentiation. Neurosci Biobehav Rev 1989; 13:301-313.
4. Morgan JI, Curran T. Stimulus-transcription coupling in neurons: role of cellular immediate-early genes. Trends Neurol Sci 1989; 12:459-462.
5. Sheng M, Greenberg ME. The regulation and function of *c-fos* and other immediate early genes in the nervous system. Neuron 1990; 4:477-485.
6. Hughes P, Dragunow M. Induction of immediate-early genes and the control of neurotransmitter-regulated gene expression within the nervous system. Pharmacol Rev 1995; 47:133-178.
7. Rauscher III FR, Cohen DR, Curran T et al. Fos-associated protein p39 is the product of the jun protooncogene. Science 1988; 240:1010-1016.
8. Kouzarides T, Ziff E. The role of the leucine zipper in the fos-jun interaction. Nature; 1988; 336:646-651.
9. Kouzarides T, Ziff E. Leucine zippers of Fos, Jun and GCN4 dictate dimerization and thereby control DNA binding. Nature 1989; 340:568-571.
10. Turner R, Tjian R. Leucine repeats and an adjacent DNA binding domain mediate the formation functional c-Fos/c-Jun heterodimers. Science 1989; 244:1689-1694.
11. Chiu R, Angel P, Karin M. JunB differs in its biological properties and is a negative regulator of c-Jun. Cell 1989; 59:979-986.
12. Suzuki T, Okuno H, Yoshida T et al. Difference in transcriptional regulatory function between c-Fos and Fra2. Nuclei Acid Res 1991; 19:5537-5542.
13. Nakabeppu Y, Nathans D. A naturally occurring truncated form of FosB that inhibits Fos/Jun transcriptional activity. Cell 1991; 64:751-759.
14. Deng TL, Karin M. JunB differs from c-Jun in its DNA-binding and dimerization domains and represses c-Jun by formation of inactive heterodimers. Genes & Dev 1993; 7:479-490.

15. Sonnenberg JL, Rauscher III FJ, Morgan JI et al. Regulation of proenkephalin by Fos and Jun. Science 1989; 246:1622-1624.
16. Meberg PJ, Gall CM, Routtenberg A. Induction of F1/GAP-43 gene: expression in hippocampal granule cells after seizures. Mol Brain Res 1993; 17:555-560.
17. Olenik C, Lais A, Meyer DK. Effects of unilateral cortex lesions on gene expression of rat cortical cholecystokinin neurons. Mol Brain Res 1991; 10:259-265.
18. Abe K, Tanzi RE, Kogure K. Selective induction of Kunitz-type protease inhibitor domain-containing amyloid precursor protein mRNA after persistent focal ischemia in rat cerebral cortex. Neurosci Lett 1991; 125; 172-174.
19. Marksteiner J, Ortler M, Bellmann R et al. Neuropeptide Y biosynthesis is markedly induced in mossy fibres during temporal lobe epilepsy of the rat. Neurosci Lett 1990; 112:143-148.
20. Lucas JJ, Mellstrom B, Colado MI et al. Molecular mechanisms of pain: serotonin$_{1A}$ receptor agonists trigger transactivation by *c-fos* of the prodynorphin gene in spinal cord neurons. Neuron 1993; 10:599-611.
21. Weiser M, Baker H, Wessel TC et al. Axotomy-induced differential gene induction in neurons of the locus ceruleus and substantia nigra. Mol Brain Res 1993; 17:319-327.
22. Goc A, Norman SA, Puchacz E et al. A 5'-flanking region of the bovine tyrosine hydroxylase gene is involved in cell-specific expression, activation of gene transcription by phorbol ester, and transactivation by c-fos and c-jun. Mol Cell Neurosci 1992; 3:383-394.
23. Hengerer B, Lindholm D, Heumann R et al. Lesion-induced increase in nerve growth factor mRNA is mediated by *c-fos*. Proc Natl Acad Sci USA 1990; 87:3899-3903.
24. Mocchetti I, De Bernardi MA, Szekely AM et al. Regulation of nerve growth factor biosynthesis by beta adrenergic receptor activation in astrocytoma cells: a potential role of c-fos protein. Proc Natl Acad Sci USA 1989; 86:3891-3895.
25. Gall CM, Isackson PJ. Limbic seizures increase neuronal production of messenger RNA for nerve-growth factor. Science 1989; 245:758-761.
26. Merlio JP, Ernfors P, Kokaia Z et al. Increased production of the TrkB protein tyrosine kinase receptor after brain insults. Neuron 1993; 10:151-164.
27. Morgan JI, Cohen DR, Hempstead JL et al. Mapping patterns of c-fos expression in the central nervous system after seizure. Science 1987; 237:192-197.
28. Robertson GS, Herrera DG, Dragunow M et al. L-dopa activates c-fos in the striatum ipsilateral to a 6-hydroxydopamine lesion of the substantia nigra. Eur J Pharmacol 1989; 159:99-100.
29. Graybiel AM, Moratalla R, Robertson HA. Amphetamine and cocaine induce drug-specific activation of the *c-fos* gene in striosome-matrix compartments and limbic subdivisions of the striatum. Proc Natl Acad Sci 1990; 87:6912-6916.

30. Hughes P, Dragunow M. Muscarinic receptor-mediated induction of Fos protein in rat brain. Neurosci Lett. 1993; 150:122-126.
31. Dragunow M, Robertson HA. Generalised seizures induce *c-fos* protein(s) in mammalian neurons. Neurosci Lett 1987; 82:157-161.
32. Morgan JI, Curran T. Proto-oncogene transcription factors and epilepsy. Trends Pharmacol Sci 1991; 12:343-349.
33. Gunn AJ, Dragunow M, Faull RLM et al. Effects of hypoxia-ischemia and seizures on neuronal and glial-like *c-fos* protein levels in the infant rat. Brain Res 1990; 531:105-116.
34. Uemura Y, Kowall NW, Flint Beal M. Global ischemia induces NMDA receptor-mediated *c-fos* expression in neurons resistant to injury in gerbil hippocampus. Brain Res 1991; 542:343-347.
35. Uemura Y, Kowall NW, Moskowitz MA. Focal ischemia in rats causes time-dependent expression of c-fos protein immunoreactivity in widespread regions of ipsilateral cortex. Brain Res 1991; 552:99-105.
36. Gass P, Spranger M, Herdegen T et al. Induction of Fos and Jun proteins after focal ischemia in the rat—differential effect of the N-methyl-D-aspartate receptor antagonist MK-801. Acta Neuropathol 1992; 84:545-553.
37. An G, Lin TN, Liu JS et al. Expression of *c-fos* and *c-jun* family genes after focal cerebral ischemia. Ann Neurol 1993; 33:457-464.
38. Gubits RM, Burke RE, Casey-MacIntosh G et al. Immediate-early gene induction after neonatal hypoxia-ischemia. Mol Brain Res 1993; 18:228-238.
39. Dragunow M, Robertson HA. Brain injury induces c-fos protein(s) in nerve and glial-like cells in adult mammalian brain. Brain Res 1988; 455:295-299.
40. Dragunow M, Goulding M, Faull RLM et al. Induction of c-fos mRNA and protein in neurons and glia after traumatic brain injury: pharmacological characterization. Exp Neurol 1990; 107:236-248.
41. Sharp JW, Sagar SM, Hisanaga K et al. The NMDA receptor mediates cortical induction of fos and fos-related antigens following cortical injury. Exp Neurol 1990; 109:323-332.
42. Herrera DG, Robertson HA. N-methyl-D-aspartate mediate activation of the *c-fos* proto-oncogene in a model of brain injury. Neurosci 1990; 35:273-281.
43. Rosen KM, McCormack MA, Villakomaroff L et al. Brief visual experience induces immediate early gene expression in the cat visual cortex. Proc Natl Acad Sci USA 1992; 89:5437-5441.
44. van der Krol AR, Mol JNM, Stuitje AR. Modulation of eukaryotic gene expression by complementary RNA or DNA sequences. Biotechniques 1988; 6:958-976.
45. Neckers L, Whitesell L. Antisense technology: biological utility and practical considerations. Am J Physiol 1993; 265:L1-L12.
46. Crooke ST. Therapeutic applications of oligonucleotide. Ann Rev Pharmacol Toxicol 1992; 32:329-376.

47. Stein CA, Tonkinson JL, Zhang LM et al. Dynamics of the internalization of phosphodiester oligodeoxynucleotides in HL60 cells. Biochemistry 1993; 32:4855-4861.
48. Yacubov LA, Deeva EA, Zarytova VF et al. Mechanism of oligonucleotide uptake by cells: involvement of specific receptors? Proc Natl Acad Sci USA 1989; 86:6454-6458.
49. Beltinger C, Saragovi HU, Smith RM et al. Binding, uptake, and intracellular trafficking of phosphorothioate-modified oligodeoxynucleotides. J Clin Invest 1995; 95:1814-1823.
50. Cazenave C, Helene C. Antisense oligonucleotides. In: Mol JNM, van der Krol AR, eds. Antisense nucleic acids and proteins: fundamental and applications. New York: M Dekker, 1991:1-6.
51. Stein CA, Tonkinson JL, Yakubov L. Phosphorothioate oligodeoxynucleotides—antisense inhibitors of gene expression? Pharmacol Rev 1991; 52:365-384.
52. Crooke ST. Progress toward oligonucleotide therapeutics: pharmacodynamic properties. FASEB J 1993; 7:533-539.
53. Ramanathan M, Macgregor RD, Hunt CA. Predictions of effect for intracellular antisense oligodeoxyribonucleotides from a kinetic model. Antisense Res Dev 1993; 3:3-18.
54. Chiasson BJ, Hooper ML, Murphy PR et al. Antisense oligonucleotide eliminates in vivo expression of *c-fos* in mammalian brain. Eur J Pharmacol 1992; 227:451-453.
55. Heilig M, Engel JA, Soderpalm B. *c-fos* antisense in the nucleus accumbens blocks the locomotor stimulant action of cocaine. Eur J Pharmacol 1993; 236:339-340.
56. Wickstrom E. Oligodeoxynucleotide stability in subcellular extracts and culture media. J Biochem Biophys Met 1986; 13:97-102.
57. Whitesell L, Geselowitz D, Chavany C et al. Stability, clearance, and disposition of intraventricularly administered oligodeoxynucleotides: implications for therapeutic applications within the central nervous system. Proc Natl Acad Sci USA 1993; 90:4665-4669.
58. Wahlestedt C. Antisense oligonucleotide strategies in neuropharmacology. Trends Pharm Sci 1994; 15:42-46.
59. Uhlmann E, Peyman A. Antisense oligonucleotides: a new therapeutic principle. Chem Rev 1990; 90:544-584.
60. Milligan JF, Matteucci MD, Martin JC. Current concepts in antisense drug design. J Med Chem 1993; 36:1923-1937.
61. Krieg AM. Uptake and efficacy of phophodiester and modified antisense oligonucleotides in primary cell cultures. Clin Chem 1993; 39:710-712.
62. Stein CA, Mori K, Loke SL et al. Phosphorothioate and normal oligodeoxyribonucleotides with 5'-linked acridine: characterization and preliminary kinetics of cellular uptake. Gene 1988; 72:333-341.
63. Katajima I, Shinohara T, Bilakovics J et al. Ablation of transplanted HTLV-I tax-transformed tumors in mice by antisense inhibtion of NF-κB. Science 1992; 258:1792-1795.

64. Thierry AR, Rahman A, Dritschilo A. Overcoming multidrug resistance in human tumor cells using free and liposomally encapsulated antisense oligonucleotides. Biochem Biophys Comm 1993; 190:952-960.
65. Pilowsky PM, Suzuki S, Minson JB. Antisense oligonucleotides: a new tool in neuroscience. Clin and Exp Pharmacol Physiol 1994; 21:935-944.
66. Leonetti JP, Degols G, Clarenc JP et al. Cell delivery and mechanisms of action of antisense oligonucleotides. Prog Nucleic Acid Res 1993; 44:143-166.
67. Ungerstedt U. Post-synaptic supersensitivity after 6-hydroxydopamine induced degeneration of the nigrostriatal dopamine system. Acta Physiol Scand Suppl 1971; 367:69-93.
68. Zigmond MJ, Abercrombie ED, Berger TW et al. Compensations after lesions of central dopaminergic neurons: some clinical and basic implications. Trends Neurosci 1990: 13:290-296.
69. Herrera-Marschitz M, Ungerstedt U. Evidence that apomorphine and pergolide induce rotation in rats by different actions on D1 and D2 receptor sites. Eur J Pharmacol 1984; 98:165-176.
70. Robertson HA, Peterson MR, Murphy K et al. D_1-dopamine receptor agonists selectively activate striatal *c-fos* independent of rotational behaviour. Brain Res 1989; 503:346-349.
71. Gerfen CR, Engber TM, Maher LC et al. D1 and D2-dopamine receptor-regulated gene expression of striatonigral and striatopallidal neurons. Science 1990; 250:1429-1432.
72. Robertson HA, Paul ML, Moratalla R et al. Expression of the immediate early gene c-fos: induction by dopaminergic drugs. Can J Neurol Sci 1991; 18 suppl 3:380-383.
73. Paul ML, Graybiel AM, David JC et al. D1-like and D2-like dopamine receptors synergistically activate rotation and *c-fos* expression in the dopamine-depleted striatum in a rat model of Parkinson's disease. J Neurosci 1992; 12:3729-3742.
74. Cole AJ, Bhat RV, Patt C et al. D1 dopamine receptor activation of multiple transcription factor genes in the rat striatum. J Neurochem 1992; 58:1420-1426.
75. Asin KE, Wirtshafter D. Effects of repeated dopamine D1 receptor stimulation on rotation and c-fos expression. Eur J Pharmacol 1993; 235:167-168.
76. Morelli M, Fenu S, Pinna A et al. Opposite effects of NMDA receptor blockade on dopaminergic D1- and D2-mediated behavior in the 6-hydroxydopamine model of turning: relationship with *c-fos* expression. J Pharmac exp Ther 1992; 260:402-408.
77. Curran T, Gordon MB, Rubino K et al. Isolation and characterization of the *c-fos* (rat) cDNA and analysis of post-translational modification in vitro. Oncogene 1987, 2:79-84.
78. Hooper ML, Chiasson BJ, Robertson HA. Infusion into the brain of an antisense oligonucleotide to the immediate-early gene *c-fos* suppresses production of Fos and produces a behavioral effect. Neurosci 1994; 63:917-924.

79. Sommer W, Bjelke B, Ganten D et al. Antisense oligonucleotide to *c-fos* induces ipsilateral rotational behaviour to d-amphetamine. NeuroReport 1993; 5:277-280.
80. Dragunow M, Lawlor P, Chiasson B et al. *c-fos* antisense generates apomorphine and amphetamine-induced rotation. NeuroReport 1993; 5:305-306.
81. Dragunow M, Tse C, Glass M et al. *c-fos* antisense reduces expression of Krox 24 in rat caudate and neocortex. Cell Mol Neurobiol 1994; 14:395-405.
82. Konradi C, Cole RL, Heckers S et al. Amphetamine regulates gene expression in rat striatum via transcription factor CREB. J Neurosci 14:5623-5634.
83. Miller JC. Induction of *c-fos* mRNA expression in rat striatum by neuroleptic drugs. J Neurochem 1990; 54:1453-1455.
84. Dragunow M, Robertson GS, Faull RLM et al. D2 dopamine receptor antagonists induce Fos and related protein(s) in rat striatal neurons. Neurosci 1990; 37:287-294.
85. Robertson GS, Vincent SR, Fibiger HC. D1 and D2 dopamine receptors differentially regulate *c-fos* expression in striatonigral and striatopallidal neurons. Neurosci 1992; 49:285-296.
86. Robertson GS, Fibiger HC. Neuroleptics increase *c-fos* expression in the forebrain: contrasting effects of haloperidol and clozapine. Neurosci 1992; 46:315-328.
87. Merchant KM, Dobie DJ, Filloux FM et al. Effects of chronic haloperidol and clozapine treatment on neurotensin and c-fos mRNA in rat neostriatal subregions. J Pharmac Exp Ther 1994; 271:460-471.
88. Kitabgi P, Carraway R, Van Rietschoten J et al. Neurotensin: specific binding to synaptic membranes from rat brain. Proc Natl Acad Sci USA. 1977; 74:1846-1850.
89. Iversen LL, Iversen SD, Bloom FE et al. Calcium-dependent release of somatostatin and neurotensin in rat brain. Nature 1978; 273:161-163.
90. Young WS III, Kuhar MJ. Neurotensin receptor localisation by light microscopic autoradiography in rat brain. Brain Res 1981; 206:273-285.
91. Uhl GR. Distribution of neurotensin and its receptors in the central nervous system. Ann NY Acad Sci 1982; 400:132-149.
92. Nemeroff CB. Neurotensin: perchance an endogenous neuroleptic? Bio Psychiat 1980; 15:283-302.
93. Govoni S, Hong JS, Yang HYT et al. Increase of neurotensin content elicited by neuroleptics in nucleus accumbens. J Pharmac Exp Ther 1980; 215:413-417.
94. Goedert M, Iversen SD, Emson PC. The effects of chronic neuroleptic treatment on neurotensin-like immunoreactivity in the rat central nervous systems. Brain Res 1985; 335:334-336.
95. Letter AA, Merchant KM, Gibb JW et al. Effect of methamphetamine on neurotensin concentrations in rat brain regions. J Pharmacol Exp Ther 1987; 241:443-447.

96. Eggerman KW, Zahm DS. Numbers of neurotensin-immunoreactive neurons selectively increased in rat ventral striatum following acute haloperidol administration. Neuropeptides 1988; 11:125-132.
97. Radke JM, McLennan AJ, Beinfield MC et al. Effects of short- and long-term haloperidol administration and withdrawl on regional brain cholecystokinin and neurotensin concentration in the rat. Brain Res 1989; 480:178-183.
98. Augood SJ, Kiyama H, Faull RLM et al. Differential effects of acute dopaminergic D1 and D2 receptor antagonists on proneurotensin mRNA expression in the rat striatum. Mol Brain Res 1991; 9:341-346.
99. Merchant KM, Miller MA, Ashleigh EA et al. Haloperidol readily increases the number of neurotensin mRNA-expressing neurons in neostriatum of the rat brain. Brain Res 1991; 540:311-314.
100. Merchant KM, Dobner PR, Dorsa DM. Differential effects of haloperidol and clozapine on neurotensin gene transcription in rat neostriatum. J Neurosci 1992; 12:652-663.
101. Merchant KM, Dobie DJ, Dorsa DM. Expression of the proneurotensin gene in the rat brain and its regulation by antipsychotic drugs. Ann NY Acad Sci 1992; 668: 54-69.
102. Merchant KM, Dorsa DM. Differential induction of neurotensin and c-fos gene expression induced by typical versus atypical antipsychotics. Proc Natl Acad Sci USA 1993; 90:3447-3451.
103. Kislauskis E, Dobner PR. Mutually dependent response elements in the cis-regulatory region of the neurotensin/neuromedin N gene integrate environmental stimuli in PC12 cells. Neuron 1990; 4:783-795.
104. Deutch AY, Lee MC, Iadarola MJ. Regionally specific events of atypical antipsychotic drugs on striatal Fos expression: the nucleus accumbens shell as a locus of antipsychotic action. Mol Cell Neurosci 1992; 3:332-341.
105. Merchant KM, Miller MA. Coexpression of neurotensin and *c-fos* mRNAs in rat neostriatal neurons following acute haloperidol. Mol Brain Res 1994; 23:271-277.
106. Nguyen TV, Kosofsky BE, Birnbaum R et al. Differential expression of *c-fos* and *zif268* in rat striatum after haloperidol, clozapine and amphetamine. Proc Natl Acad Sci USA 1992; 89:4270-4274.
107. Rogue P, Vincendon G. Dopamine D2 receptor antagonists induced immediate early genes in the rat striatum. Brain Res Bull 1992; 29:469-472.
108. Merchant KM. *c-fos* antisense oligonucleotide specifically attenuates haloperidol-induced increases in neurotensin/neuromedin-N messenger RNA expression in rat dorsal striatum. Mol Cell Neurosci 1994; 5:336-344.
109. Robertson GS, Tetzlaff W, Bedard A et al. *c-fos* mediates antipsychotic-induced neurotensin gene expression in the rodent striatum. Neurosci 1995; 67:325-344.
110. Konradi C, Kobierski LA, Nguyen TV et al. The cAMP-response-element-binding protein interacts but Fos protein does not interact, with the proenkephalin enhancer in rat striatum. Proc Natl Acad Sci USA 1993; 90:7005-7009.

111. Konradi C, Heckers S. Haloperidol-induced Fos expression in striatum is dependent upon transcription factor cyclic AMP response element binding protein. Neurosci 1995; 65:1051-1061.
112. Chalmers J, Arnolda L, Llewellyn-Smith I et al. Central neurons and neurotransmitters in the control of blood pressure. Clin Exp Pharmacol Physiol 1994; 21:819-829.
113. McKitrick DJ, Krukoff TL, Calaresu FR. Expression of c-fos protein in rat brain after electrical stimulation of the aortic depressor nerve. Brain Res 1992; 599:215-222.
114. Badoer E, McKinley MJ, Oldfield BJ et al. Distribution of hypothalamic, medullary and lamina terminalis neurons expressing Fos after haemorrhage in conscious rats. Brain Res 1992; 582:323-328.
115. Dun NJ, Dun SL, Chiaia NL. Haemorrhage induces Fos immunoreactivity in rat medullary catecholaminergic neurons. Brain Res 1993; 608:223-232.
116. Chiu TH, Dun SL, Tang H et al. c-fos antisense attenuates Fos expression in rat central neurons induced by haemorrhage. NeuroReport 1994; 195:2178-2180.
117. Barman SM, Gebber GL. Axonal projection patterns of ventrolateral medullospinal sympathoexcitatory neurons. J Neurophysiol 1985; 53:1551-1566.
118. Blessing WW. Depressor neurons in rabbit caudal medulla act via GABA receptors in rostral medulla. Am J Physiol 1988; 254 (Heart Circ Physiol 23):H686-H692. see text sentence doesnt make sense
119. Chalmers JP, Pilowsky PM. Brainstem and bulbospinal neurotransmitter systems in the control of blood pressure. J Hyperten 1991; 9:675-694.
120. Guyenet PG, Haselton JR, Sun MK. Sympathoexcitatory neurons of the rostroventrolateral medulla and the origin of the sympathetic vasomotor tone. Prog Brain Res 1989; 81:105-116.
121. Minson JB, Suzuki S, Llewellynsmith IJ et al. c-fos expression in central cardiovascular pathways. Clin Exp Hypertens 1995; 17:67-79.
122. Minson JB, Suzuki S, Llewellynsmith IJ et al. Disinhibition of the rostral medulla increases blood pressure and Fos expression in bulbospinal neurons. Brain Res 1994; 646:44-52.
123. Suzuki S, Pilowsky P, Minson J et al. c-fos antisense in rostral ventral medulla reduces arterial blood pressure. Am J Physiol 1994; 266: R1418-R1422.
124. Takahashi JS. Circadian-clock regulation of gene expression. Curr Opin Genet Dev 1993; 3:301-309.
125. Rea MA. Light increases fos-related protein immunoreactivity in the rat suprachiasmatic nuclei. Brain Res Bull 1989; 23:577-581.
126. Aronin N, Sagar SM, Sharp FR et al. Light regulates expression of a fos-related protein in rat suprachiasmatic nucleus. Proc Nat Acad Sci USA 1990; 87:5959-5962.
127. Rusak B, Robertson HA, Wisden et al. Light pulses that shift rhythms induce gene expression in the suprachiasmatic nucleus. Science 1990; 248:1237-1240.

128. Earnest DJ, Iadarola M, Yeh HH et al. Photic regulation of *c-fos* expression in neural components governing the entrainment of circadian rhythm. Exp Neurol 1990; 109:353-361.
129. Kornhauser JM, Nelson DE, Mayo KE et al. Photic and circadian regulation of *c-fos* gene expression in the hamster suprachiasmatic nucleus. Neuron 1990; 5:127-134.
130. Rusak B, McNaughton L, Robertson HA et al. Circadian variation in photic regulation of immediate early gene mRNAs in rat suprachiasmatic nucleus cells. Mol Brain Res 1992; 14:124-130.
131. Sutin EL, Kilduff TD. Circadian and light-induced expression of immediate-early gene mRNAs in the rat suprachiasmatic nucleus. Mol Brain Res 1992; 15:281-290.
132. Wollnik F, Brysch W, Uhlmann E et al. Block of c-Fos and Jun B expression by antisense oligonucleotides inhibits light-induced phase shifts of the mammalian circadian clock. Eur J Neurosci 1995; 7:388-393.
133. Gillardon F, Beck H, Uhlmann E et al. Inhibition of c-fos protein expression in rat spinal cord by antisense oligodeoxynucleotide superfusion. Eur J Neurosci 1994; 6:880-884.
134. Wollnik F. Strain differences in the pattern and intensity of wheel running activity in laboratory rats. Experientia 1991; 47:593-598.
135. McGinty D, Szymusiak R. Keeping cool: a hypothesis about the mechanisms and functions of slow-wave sleep. Trends Neurosci 1991; 13:480-487.
136. Pompeiano M, Cirelli C, Tonini G. Effects of sleep deprivation on Fos-like immunoreactivity in the rat brain. Arch Ital Biol 1992; 130:325-335.
137. O'Hara BF, Young KA, Watson FL et al. Immediate-early gene expression in brain during sleep deprivation: preliminary observations. Sleep 1993; 16:1-7.
138. Cirelli C, Pompeiano M, Arrighi P et al. Sleep-waking changes after *c-fos* antisense injections in the medial preoptic area. NeuroReport 1995; 6:801-805.
139. Herdegen T, Kovary K, Leah J et al. Specific temporal and spatial distribution of JUN, FOS and KROX-24 proteins in spinal neurons following noxious transsynaptic stimulation. J Comp Neurol 1991; 313:178-191.
140. Hunt SP, Pini A, Evan G. Induction of c-fos-like protein in spinal cord neurons following sensory stimulation. Nature 1987; 328:632-634.
141. Draisci G, Iadarola MJ. Temporal analysis of increases in *c-fos*, preprodynophin and preproenkephalin mRNAs in rat spinal cord. Mol Brain Res 1989; 6:31-37.
142. Wisden W, Errington ML, Williams S et al. Differential expression of immediate early genes in the hippocampus and spinal cord. Neuron 1990; 4:603-614.
143. Abbadie C, Lombard MC, Morain F et al. Fos-like immunoreactivity in the rat superficial dorsal horn induced by formalin injection in the forepaw: effect of dorsal rhizotomies. Brain Res 1992; 578:17-25.

144. Pettersson EKE, Hunter J, Poat JA et al. The effect of enadoline on immediate early gene expression in the rat paw formalin test. Br J Pharmac 1994; 110:61P.
145. Draisci G, Kajander KC, Dubner R et al. Up-regulation of opioid gene expression in spinal cord evoked by experimental nerve injuries and inflammation. Brain Res 1991; 560:186-192.
146. Noguchi K, Kowalski K, Traub R et al. Dynorphin expression and Fos-like immunoreactivity following inflammation induced hyperalgesia are colocalized in spinal cord neurons. Mol Brain Res 1991; 10:227-233.
147. Cesselin F, Montastruc JL, Gros C et al. Met-enkephalin levels and opiate receptors in the spinal cord of chronic suffering rats. Brain Res 1980; 191:289-293.
148. Allerton CA, Smith JAM, Hunter JC et al. Correlation on ontogeny with function of [3H]-U69593 labelled k-opioid binding sites in the rat spinal cord. Brain Res 1989; 502:149-157.
149. Yoshimura M, North RA. Substantia gelatinose neurons hyperpolarised in vitro by enkephalin. Nature 1983; 305:529-530.
150. Millan MJ. Kappa opioid receptors and analgesia. Trends Pharmac Sci 1990; 11:70-76.
151. Civelli O, Douglass J, Goldstein A et al. Sequence and expression of the rat prodynophin gene. Proc Natl Acad Sci USA 1985; 82:4291-4295.
152. Hunter JC, Woodburn VL, Durieux C et al. *c-fos* antisense oligodeoxynucleotide increases formalin-induced nociception and regulates preprodynorphin expression. Neurosci 1995; 65:485-492.
153. Hughes P, Singleton K, Dragunow M. MK801 does not attenuate immediate-early gene expression following an amygdala afterdischarge. Exp Neurol 1994; 128:276-282.
154. Abe K, Tanzi RE, Kogure K. Induction of HSP70 and mRNA after transient global ischemia in gerbil brain. Neurosci Lett 1991; 125:66-71.
155. Blumenfield KS, Welsh FA, Harris VA et al. Regional expression of *c-fos* and heat shock protein-70 mRNA following hypoxia-ischemia in immature rat brain. J Cereb Blood Flow Metab 1992; 12:987-995.
156. Jorgensen MB, Deckert J, Wright DC et al. Delayed *c-fos* proto-oncogene expression in the rat hippocampus induced by transient global cerebral ischemia: an in situ hybridization study. Brain Res 1989; 484:393-398.
157. Onodera H, Kogure K, Ono Y et al. Proto-oncogene *c-fos* is transiently induced in the rat cerebral cortex after forebrain ischemia. Neurosci Lett 1989; 98; 101-104.
158. Wessel TC, Joh TH, Volpe BT. In situ hybridisation analysis of *c-fos* and c-jun expression in the rat brain following transient forebrain ischemia. Brain Res 1991; 567:231-240.
159. Nowak TS Jr, Ikeda J, Nakajima T. 70-kDa heat shock protein and *c-fos* gene expression after transient ischaemia. Stroke 1990; 21:107-111.
160. Sharp FR, Lowenstein D, Simon R et al. Heat shock protein hsp72 induction in cortical striatal astrocytes and neurons following infarction. J Cereb Blood Flow Metab 1991; 11:621-627.

161. Vass K, Welck WJ, Nowak Jr TS. Localization of 70-kDa stress protein induction in gerbil brain after ischemia. Acta Neuropathol 1988; 77:128-132.
162. Lindvall O, Ernfors P, Bengzon et al. Differential regulation of mRNAs for nerve growth factor, brain-derived neurotrophic factor, and neurotrophin 3 in the adult rat brain following cerebral ischemia and hypoglycemic coma. Proc Natl Acad Sci USA 1992; 89:648-652.
163. Tominaga T, Kure S, Narisawa K et al. Endonuclease activation following focal ischemic injury in the rat brain. Brain Res 1993; 608:21-26.
164. Shigeno T, Mima T, Takakura K et al. Amelioration of delayed neuronal death in the hippocampus by nerve growth factor. J Neurosci 1990; 11:2914-2919.
165. Goto K, Ishige A, Sekiguchi K et al. Effects of cycloheximide on delayed neuronal death in rat hippocampus. Brain Res 1990; 534:299-302.
166. Pappas S, Crepel V, Hasboun D et al. Cycloheximide reduces the effects of anoxic insult in vivo and in vitro. Eur J Neurosci 1992; 4:758-765.
167. Liu PK, Salminen, A, He YY et al. Suppression of ischemia-induced fos expression and AP-1 activity by an antisense oligodeoxynucleotide to *c-fos* mRNA. Ann Neurol 1994; 36:566-576.

CHAPTER 9

ANTISENSE STRATEGY AND STIMULUS-SECRETION COUPLING

Marjan Rupnik and Robert Zorec

INTRODUCTION

The term stimulus secretion-coupling was introduced by Douglas and Rubin[1] and describes a chain of events that leads from stimulus (occupancy of a receptor, or voltage modulation of membrane conductances) to activation of intracellular signaling pathways which results in the modulation of hormone or neurotransmitter secretion. With anterior pituitary cells, the stimulus signal is typically a small blood-borne hypothalamic peptide (a "releasing factor"), whereas in neurones the stimulus of neurotransmitter release is the appearance of an action potential at the nerve terminal.

It has been thought for more than 30 years that the stimulus leads to a change in cytosolic calcium homeostasis leading to augmented exocytosis of vesicles or secretory granules with plasmalemma. This primary response may be modulated at two levels: (i) at the site of calcium entry; and (ii) at the site of exocytosis—but detailed knowledge about the molecular events is fragmental. The major obstacle in elucidating these events is the relative inaccessibility of the cytosol, especially if we consider studying responses at cellular level. One promising approach in this quest is to employ new methods, such as the antisense strategy to inhibit expression of specific proteins putatively playing a role in stimulus-secretion coupling. The aim of this chapter is to illustrate the application of antisense oligodeoxyribonucleotides (oligonucleotides) to study the physiological role of specific proteins in stimulus-secretion coupling. This strategy is still in the early stage mainly confounded by the techniques of antisense oligonucleotide delivery into cells. (Also see ref. 2.)

MECHANISM(S) OF ACTION OF ANTISENSE OLIGONUCLEOTIDES ON PROTEIN EXPRESSION

For antisense oligonucleotides to block the chain leading to protein expression, these substances have to be internalized. This presents a problem since oligonucleotides are highly charged molecules and plasmamembrane is a barrier for diffusion. In addition, nucleases in serum degrade oligonucleotides. Thus a relatively high concentration (about 10 µM) is used to achieve internalization of a small fraction of extracellular pool of molecules. Moreover such relatively high concentrations may be toxic to some cells,[3] and the cost of experiments is also considerable. A number of modifications of oligodeoxyribonucleotide chemical structure and/or conjugation to transport vectors have been tested to reduce these problems (see below).[4]

Once inside the cell oligonucleotides must escape intracellular nucleases and hybridize with the targeted sequence (and not to any other). The inhibitory effect of antisense oligonucleotides may occur in the nucleus or in the cytoplasm, at a number of steps of protein expression (see Fig. 9.1). The spectrum of actions of antisense oligonucleotides is not totally established.[5] In physiology oligonucleotides are designed to interact mainly with messenger RNA (mRNA). Translation

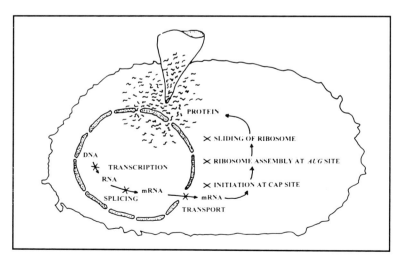

Fig. 9.1. Interrupting the chain of protein expression. Crosses (X) indicate the possible sites of action of the antisense oligodeoxynucleotides. Impalement by a microelectrode indicates a method to introduce oligodeoxynucleotides into cytoplasm and nucleus.

machinery arrest results from a physical block to binding of the initiation complex at the AUG initiation codon or at the region that covers the ribosome binding site, upstream of the initiation codon (see Fig. 9.1). Binding of oligonucleotides may also have two other consequences; an increase in the catabolic rate of the targeted mRNA due to a change of its secondary structure or the arrest of exportation of the mRNA from the nucleus. Another mechanism of action of antisense oligonucleotides is the activation of RNAse H, an ubiquitous enzyme found both in cytoplasm and in nucleus[6] that cleaves the target RNA-DNA oligonucleotide duplexes. Several chemically modified oligodeoxynucleotides are unable to modify RNAse H activity, such as methylphosphonates[7] and α-oligomers,[8] and require higher effective concentrations.[4]

Several points have to be considered before using antisense oligonucleotides in experiments. For a review see refs. 3, 9-14. One of this points is selectivity of oligonucleotides. The specificity of an antisense oligonucleotide depends on its length. The longer an oligonucleotide is, the more likely it is to bind to one and only one DNA or RNA target; however longer oligonucleotides may not be specific due to recognition of slightly mismatched sequences. A minimum level of affinity is required for the desired specific interaction and this can be achieved with an oligonucleotide that is 11 to 15 nucleotides in length.[3] To help predict antisense oligonucleotide efficacy, computational approaches have been designed (see refs. 15, 16); however, it is imperative that every oligonucleotide sequence is checked for similarities with other sequences present in gene databases.

Stability of these compounds has to be considered as well since the half-life of the oligonucleotide probe should exceed that of the targeted protein to be effective. Figures 9.2 and 9.4 show examples of a typical transient effect of applied antisense probes on responses of clonal cells. The transient nature of antisense oligonucleotide treatment is due to the limited stability of the probe. The stability of the duplex increases with oligomer length.[17] Moreover, a wide range of modifications of the oligonucleotide backbone have been introduced to increase stability.[4] For example, phosphorothioate oligodeoxyribonucleotide probes (sulphur atoms substituted for oxygen atoms in the sugar phosphate backbone[18]) have half lives of more than 24 hours in serum media,[3] but their intracellular stability was not determined.[19] There are also nonionic methylphosphonate analogs which are also extremely stable to nucleases, but more easily enter the cytosol in comparison to the charged oligonucleotides. Furthermore there is a novel class of chimeric oligonucleotides, consisting of mixed modifications: for example of some unmodified nucleotides flanked at 3' and 5' parts and of some methylphosphonate nucleotides.[20] Such chimeric oligonucleotides combine the advantages of methylphosphonates (improved uptake, nuclease resistance) and of the unmodified oligonucleotides

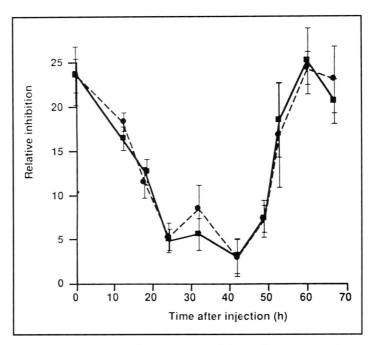

Fig. 9.2. Time course of Ca^{2+}-current inhibition by somatostatin or carbachol in GH_3 cells. At time zero cells were injected with antisense oligonucleotide against all β-subunits of heterotrimeric GTP-binding proteins (5'-TTGCAGTTGAAGTCGTCRTA-3'). At indicated time points, the Ca^{2+} current was measured in the presence of 1μM somatostatin (dotted line) or 10 μM carbachol (solid line). Mean values with s.e.m. are shown (N = > 5). Reprinted with permission from: Kleuss et al. Nature 1991; 358:424-426, Copyright 1995 Macmillan Magazines Limited.

(RNAse H cleavage of target mRNA, improved water solubility). The disadvantage of these new chimeric oligonucleotides is their relatively high price,[4] and applications with these probes in physiology are still at their infancy.

A major advantage of the antisense oligonucleotide approach is that the simple synthesis and testing of a number of oligonucleotides enables for a screening process before their use. A "trial and error" approach must be performed to determine whether translated mRNA regions (including among them the initiation codon) as well as 5'- (e.g., capping region) or 3'-untranslated messenger RNA regions can be targeted. Most antisense oligonucleotides (nonmodified) have been designed to bind to the translation initiation codon or its immediate vicinity after having tested them in vitro.[21] Oligonucleotides designed to bind to the capping region have been used as well.[22]

The antisense oligonucleotide strategy has clearly a number of important advantages, but there are also disadvantages.[13] Apart from

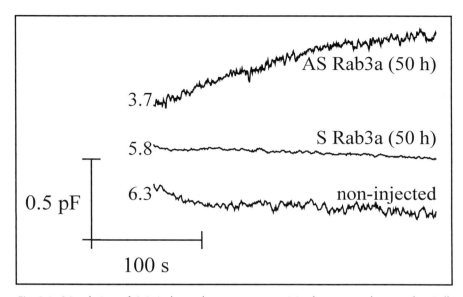

Fig. 9.3. Stimulation of Ca^{2+}-independent secretory activity from rat melanotrophs. Cells were microinjected with antisense (AS Rab 3a) and sense (S Rab 3a) oligodeoxynucleotides directed against mRNA encoding Rab3a protein. Membrane capacitance (C_m) was monitored using the whole-cell patch-clamp technique 50 hours after the injection of the probes indicated. Pipette solution contained 4 mM EGTA[74] and was essentially calcium-free. In these conditions control cells responded with a steady decrease in C_m (-2.5 ±0.8 %, n = 31), measured 200 s after the establishment of whole-cell recording. Fifty hours incubation of cells with the sense Rab3a probe had no effect on the cellular response (-0.8 ± 3.0 %, n = 6). In contrast, antisense Rab3a treated cells responded with a steady increase in C_m, significantly different from noninjected control and sense control (6.4 ± 0.4 %, n = 8; Student t-test, $p < 0.001$). Ninety hours after microinjection in both antisense and sense treated cells C_m was not changing or it decreased when measured in a whole-cell mode (antisense, 0.1 ± 1.2 %, n = 12; sense, -2.4 ± 0.2 %, n = 3). Sequences of oligunucleotides used: antisense Rab3a, 5'-TGT GGC TGA GGC CAT CTT GCC C-3'; sense Rab3a, 5'-ACC ATC TAC CGA AAT GAC AAG-3'.[6]

occasional nonspecific or toxic actions, the greatest drawback may be incomplete effect of oligonucleotide treatment. For example, if a biological function requires only few protein molecules for a full response, but in the cell these are synthesized redundantly, oligonucleotide treatment may be "functionally" silent.

DELIVERY OF ANTISENSE OLIGONUCLEOTIDES INTO CYTOSOL

Currently the most popular forms of oligonucleotides used in physiology are the unmodified (phosphodiester) and the phosphorothioate oligodeoxyribonucleotides. Because the target of these molecules is in the cytosol/nucleus and the membrane is a barrier, several techniques of delivery were introduced, such as cellular uptake with various vectors and conjugates, membrane permeabilization techniques, loading cells by whole-cell recording mode of patch-clamp and loading by microinjection (Fig. 9.3).

CELLULAR UPTAKE OF OLIGONUCLEOTIDES AND VECTOR/CONJUGATE SYSTEMS

The easiest way to deliver antisense oligonucleotides into cytosol of a cell population is to soak the cells with the solution containing these probes. Diffusion of molecules across plasmalemma is insignificant and the main pathway of oligonucleotide uptake proceeds by endocytosis in a temperature-dependent, saturable fashion.[23] In addition, oligonucleotides taken up by the cell, may not be compartmentalized properly for the antisense activity.[24] Unlike viruses, which use membrane fusion to deliver nucleic acids to the cytoplasm, most oligonucleotides remain trapped in the vesicles, where they are degraded in the lysosomes or are simply returned to the extracellular medium.[23]

Another problem with such a delivery of oligonucleotides is their sensitivity to degradation by nucleases present in the serum, especially 3'-5' exonucleases. The half-life of unmodified oligonucleotides in the presence of serum can be as low as 15 minutes.[25]

At least two solutions were employed to avoid poor absorption and degradation. Firstly, as discussed above oligonucleotides can be chemically modified to increase their cellular uptake and nuclease resistivity. Secondly, a number of vectors/conjugates linked to oligonucleotides were tested. For example, increased cytoplasmic delivery can be achieved by conjugation of oligonucleotides to cholesterol. In addition to enhanced uptake of probes with this modification, unfortunately, also the toxicity of the conjugate is increased.[23]

Another strategy of antisense oligonucleotide delivery consists in encapsulating either unmodified or modified oligonucleotides in liposomes.[23,26,27] Liposomes protect the oligonucleotides from exonucleases, and enhance their intracellular uptake up to an order of magnitude.[28] Positively charged groups of cationic lipids interact directly with the negatively charged phosphate residues of the oligonucleotide.[29] On the other hand the success of oligonucleotide uptake by fusion of cells with liposomes depends on the specific system in use.[26] Another potential drawback of this strategy includes cellular toxicity and decreased efficiency in serum-containing media.[4]

For the purpose of a putative antisense therapy more complex constructs were attempted, such as the condensation of DNA with transferrin receptor and endosome lytic peptides or adenovirus.[14] Transferrin receptor is present on most proliferating and some nonproliferating higher eukaryotic cells as well as cancer cells. A more specific delivery system to be used as a putative anti-cancer system consists of liposomes and antisense oligonucleotides conjugated to folate, since predominantly cancer cells express folic acid receptors.[30]

Finally, one should consider that the delivery of antisense oligonucleotides into all cells may not be equal by any of the aforementioned techniques. The need for a stable molar excess of antisense probes over the endogenous mRNA in nearly 100% of cells in the correct subcellular compartment may exceed the potential of the described methods, if used in single cell physiology, but future work will provide an answer.

PERMEABILIZATION OF PLASMAMEMBRANE FACILITATES OLIGONUCLEOTIDE DELIVERY

To introduce relatively large charged molecules into cytosol one can employ many permeabilization techniques, which allow the selective poration of the plasma membrane while maintaining their general structural integrity. The most widely applied permeabilization techniques to study stimulus-secretion coupling include the application of bacterial cytolysins, streptolysin-O[31,32] and streptococcal α-toxin,[33] or application of plant glycosides such as saponin[34] and digitonin.[35] Also high voltage discharges may be used,[36] and in some susceptible cell types, a tetrabasic free acid of ATP (ATP^{4-})[37] was used to porate the plasma membrane. Distinct features, including the nature of the damage inflicted, duration and pore dimensions, of mentioned permeabilization techniques present a substantial repertoire in designing the experimental strategies.[38]

Recently, a streptolysin-O (SLO) based permeabilization was described as a simple, highly efficient introduction of oligonucleotides into eukaryotic cells, devoid of measurable toxicity.[32] Nonmodified antisense oligonucleotides can be used in such experiments. Streptolysin-O binds to cholesterol in the plasma membranes and subsequently assembles into large supramolecular pores that penetrate and span the lipid bilayer.[31] Since the cholesterol content may vary among the different cell types, the success of the streptolysin-O aided oligonucleotide delivery depends on the optimization of permeabilization conditions.[32] We have adopted a similar permeabilization approach to study the role of heterotrimeric GTP-binding proteins in secretory activity of rat basophilic leukemia cells (Fig. 9.4). Monolayers of cultured cells were transiently permeabilized with streptolysin-O in the oligonucleotide rich extracellular cell culture solution. Washing away the permeabilizing agent led to the recovery, while the antisense probes were trapped in the cytosol. Another promising antisense oligonucleotide delivery system could be the permeabilization by ATP.[39]

Permeabilization techniques to introduce antisense oligonucleotides into cytosol of cell populations have an advantage in comparison to the simple cellular uptake method since lower concentrations of oligonucleotides are required, and delivery into all permeabilized cells is nearly homogeneous.

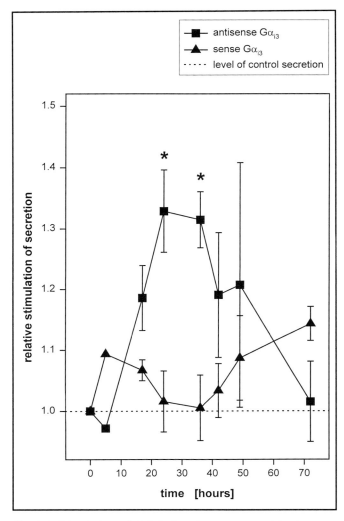

Fig. 9.4. Stimulation of Ca^{2+}-induced secretory activity from rat basophilic leukemia cells. Cells were SLO permeabilized in the presence of antisense and sense oligonucleotides directed against mRNA encoding α_{i3} subunits (see chapter 3.2.). Hexosaminidase level was measured 10 minutes after the cells were stimulated with ionomycin (2.5 µM) at time points between 5 and 72 hours after the oligonucleotide treatment. Sequences of oligonucleotides used: antisense $G\alpha_{i3}$, 5'-CAT GAC GGC CGG AGA GG-3';[75] sense $G\alpha_{i3}$, 5'-CCT CTC CGG CCG TCA TG-3'. The stimulation of Ca^{2+}-induced secretory activity is shown as an index of secretory activity in control cells (no oligodeoxynucleotides added). Mean ± standard error of the mean is shown, number of tested cell populations was from 2 to 5. Asterisks indicate time points where the difference between antisense and sense treatment was significant (Student t-test, $p < 0.005$).

Loading Antisense Oligonucleotides with the Whole-Cell Patch Clamp Technique and Microinjection Techniques

In single cell studies antisense oligonucleotides are introduced into the cytosol most conveniently either by a patch-pipette or by a micropipette with a pressure ejection system. While the latter systems can be purchased also at a substantial cost, a standard patch-clamp electrophysiological laboratory can be successfully employed at no additional cost to deliver antisense oligonucleotides into single cells.

The patch-clamp techniques[40] provide many possibilities to study currents through single channels and through the whole-cell membrane.[41] Moreover, changes in surface area of cells due to exo- and endocytosis can be monitored by whole-cell membrane capacitance measurements. Electrical capacity of cell membrane (C_m) is linearly related to the surface of the membrane.[42] An additional advantage of the whole-cell patch-clamp technique is also that it provides means to introduce substances into the cytosol. Molecules, ions, peptides, toxins, antibodies, oligonucleotides, putatively interfering with cytosolic processes can be dissolved in the pipette-filling solution, and upon the formation of the whole-cell configuration the diffusional barrier between pipette lumen and cytosol is eliminated, allowing an exchange of substances between cytosol and pipette lumen.[41,43] However, a word of caution is required. Loading large quantities of high-molecular weight substances (such as antibodies) may be impractical since cellular constituents can be lost into the pipette. Nonetheless, antisense oligonucleotides (about 6 to 7 kDa) are of just the right size to be introduced with this technique.[2] With this technique the quality of loading can be monitored, and the number of molecules entered the cellular interior estimated.[43]

A somehow different approach of delivery of antisense oligonucleotides into the cell interior in single cell studies is microinjection. In contrast to the patch-clamp technique, very sharp micropipettes are used and for a small volume of concentrated content to be introduced into cells, pressure ejection is used. Usually, injection is monitored by coinjecting a fluorescent marker (preferably of the same molecular weight), which can be used later to identify injected cells.[44] Some workers were injecting antisense oligonucleotide probes directly into the nucleus.[45] However, injection of antisense oligonucleotides into cytosol may in the end reach putative targets in the nucleus as well. Fluorescently labeled oligonucleotides injected into the cytosol were also found to stain the nucleus.[46] In comparison with patch-clamp loading the advantage of microinjecting antisense oligonucleotides is that one micropipette can be used to load more than one cell, and endogenous cellular compounds are retained during such a procedure.[47] The disadvantage of the technique is that the method is technically demanding and requires specialized equipment.

APPLICATION OF ANTISENSE STRATEGY TO STUDY THE MODULATION AND FUNCTION OF VOLTAGE-ACTIVATED CALCIUM CHANNELS

In contrast to nonexcitable cells where the activation of GTP-binding proteins appears to be the crucial step to trigger secretion, in excitable cells a rise in cytosolic calcium is required.[48] Thus the entry of calcium ions through voltage-activated calcium channels (VACC) is an important control step in the chain of events leading from stimulus to secretion.

The activity of VACC, composed of α_1, α_2-δ, β and γ subunits, is modulated by numerous hormones and neurotransmitters interacting with membrane receptors. Most of modulation mechanisms involve GTP-binding proteins.[49] The mechanism of ion channel modulation by GTP-binding proteins may be direct with a given channel in a "membrane delimited" manner. On the other hand GTP-binding proteins may also activate an intracellular pathway which secondarily modifies the parameters of ion channel function.[49,50,51] To understand the molecular events in these mechanisms we need to identify different GTP-binding proteins involved in this signaling pathways. However, the multiplicity of GTP-binding proteins,[52] as well as VACC and their subunits,[49] makes this a very difficult task.

In excitable endocrine cells such as in lactotroph anterior pituitary cells, the activity of voltage-activated channels is controlled by various release factors including dopamine.[53] Dopamine receptors are seven transmembrane-segment proteins, and so they are believed to activate heterotrimeric GTP-binding proteins in order to achieve their effect on ion channels. The question is which of the molecular forms of GTP-binding proteins mediate these effects. In patch-clamp loaded normal lactotrophs with antisense oligonucleotides to block the expression of α_o, α_{i1}, α_{i2} and α_{i3} it was shown that dopamine modulates VACCs via $\alpha_o\beta\gamma$ heterotrimeric GTP-binding proteins.[53]

Kleuss et al[22,45,54] injected the nucleus of clonal pituitary cells with unmodified oligonucleotides targeted against α_o, α_i, β or γ-subunits of the heterotrimeric GTP-binding proteins. Calcium current responses to somatostatin and carbachol were recorded 2-3 days later in the microinjected cells when responses were maximal (see Fig. 9.2). The results suggested that the VACC is coupled through a $\alpha_{o1}\beta_3\gamma_4$ heterotrimeric GTP-binding protein to the muscarinic receptor, and via a $\alpha_{o2}\beta_1\gamma_3$ heterotrimeric GTP-binding protein to the somatostatin receptor. Whereas α_{i2} of the heterotrimeric GTP-binding proteins is required for thyrotropin (TRH) stimulation of VACCs in clonal pituitary cells.[55] Similarly, in sensory neurones the GABA$_B$ receptor modulation of VACCs was found to be mediated via the α_o heterotrimeric GTP-binding protein.[56] These experiments clearly illustrate that different receptors in a cell type modulate the activity of VACCs via different heterotrimeric GTP-binding proteins.

Antisense strategy can also be employed to study the role of different VACC subunits. The effect of the β-subunit depletion on pharmacological and biophysical properties of VACCs was studied in cultured dorsal root ganglion (DRG) neurons.[57] A 26 mer antisense oligonucleotides with homology to all published VACC β-subunits was microinjected into cells and maximal depletion of VACC β-subunit immunoreactivity was observed after about 180 hours, suggesting a half-life of turnover of this subunit of more than 50 hours. In depleted cells maximal current was attenuated by about 50% with a shift in voltage dependence of current activation of about +7 mV. The ability of a 1,4-dihydropyridine agonist to enhance calcium current was greatly reduced following depletion of VACC β-subunit. Since there are four VACC β-subunits, with many splice variants,[58] it will be interesting to see which of the VACC β-subunits are responsible for observed effects in DRG neurones.[57] In neuroblastoma cells the inhibition of expression of the $β_2$-subunit by antisense oligonucleotides increases the surface expression of the $α_1$ subunit of N-type ωCTX sensitive VACC, suggesting that the $β_2$-subunit is involved in the transport of $α_1$ subunits to plasmalemma.[59] Antisense oligonucleotides against the $α_1$ subunit (rbA, P-type) VACCs were used to inhibit expression of these channels in an oocyte system.[60]

All these examples demonstrate the potential of antisense strategy both in studies where the modulation of VACCs by GTP-binding proteins and the role of channel subunits is studied.

ANTISENSE STRATEGY AND THE ASSIGNMENT OF MONOMERIC AND HETEROTRIMERIC GTP-BINDING PROTEINS IN THE SECRETORY ACTIVITY

Among many recently identified proteins evidence is accumulating that both monomeric and heterotrimeric GTP-binding proteins are involved in regulated exocytosis.[21,38,61–68]

An unequivocal role for monomeric GTP-binding proteins in calcium dependent exocytosis was demonstrated by depleting Rab3b protein by loading single anterior pituitary cells with antisense oligonucleotides with the patch-clamp technique.[21] Loaded cells were stimulated 48 hours later by cytosol dialysis with a high calcium containing pipette solution, while the secretory response was monitored by measurements of membrane capacitance, a parameter related to surface area.[42] Depletion of Rab3b, but not Rab3a isomer resulted in an attenuation of the Ca^{2+}-induced secretory activity. In contrast in chromaffin cells Rab3a protein was shown to be involved in Ca^{2+}-dependent secretory activity.[69]

In some anterior pituitary cells and in particular in the rat pars intermedia Rab3a and Rab3b appear to be colocalized.[70] Thus a question arises whether both proteins may share the same function or are involved in distinct processes. Figure 9.3 shows an experiment where single melanotrophs from rat pars intermedia were microinjected with antisense oligonucleotides against the Rab3a.[21] Injection with antisense, but not the sense probe, resulted in a Ca^{2+}-independent rise in membrane surface area, indicating a role of Rab3a in membrane turnover. Thus in melanotrophs Rab3b is involved in the regulation of Ca^{2+}-dependent (not shown)[21] and Rab3a in Ca^{2+}-independent membrane turnover phenomena. In the future it will be imperative to elucidate whether both Rab3 isomers control the pathway leading to exocytosis, as is the case for Rab3b.[21]

We also employed the antisense strategy to demonstrate a role for heterotrimeric GTP-binding proteins in the regulation of Ca^{2+}-dependent release of hexosaminidase in rat basophilic leukemia cells (RBL). A role of these proteins in secretory activity of RBL cells was suggested previously.[71] Rat basophilic leukemia cells were found to be a stable cell line for the biochemical studies of secretion of a cell population. Antisense and sense oligonucleotides directed towards mRNA coding for different Gα subunits were delivered into cells by a transient streptolysin-O permeabilization (Fig. 9.4). Cells were left to recover for several hours and were then stimulated with ionomycin to secrete.[72] Streptolysin-O permeabilization was reversible only if applied to cells firmly attached to the bottom of a flat 300 µl well. Hexosaminidase level was subsequently essayed as a measure of secretory activity.[73] Cells attached to the bottom of one well yielded enough hexosaminidase for differential analysis. Also ionomycin stimulation was reversible allowing several stimulations of the same population of cells during a period from 6 to 72 hours after the reversible streptolysin-O permeabilization. As shown on Figure 9.4 at time 0 cells were permeabilized with streptolysin-O in the presence of oligonucleotides (10 µM). The effect of antisense and sense probes were determined. As in microinjected cells (Fig. 9.2) the effect of antisense oligonucleotide was transient (Fig. 9.4) with a maximum effect between 25 to 40 hours after the loading. Stimulated cells with ionomycin after 24 hours secreted significantly more than in control conditions, if they were loaded with antisense oligonucleotides against α_{i3}. Cells loaded with sense oligonucleotides did show a small insignificant increase in hexasominidase secretion, probably due to a nonspecific effect. After 72 hours ionomycin stimulation did not reveal any significant differences between the nonoligonucleotide control and antisense or sense oligonucleotide treated cells. It can be concluded that α_{i3} heterotrimeric GTP-binding protein is inhibitory to secretory activity in RBL cells. These results contrast to recent findings about the role of a heterotrimeric GTP-binding protein in mast cells based on the synthetic peptide and antibody studies.[66] The differences may be due to different techniques employed and cell types studied.

SUMMARY AND PERSPECTIVES

The antisense strategy is a relatively new approach in single cell physiology. Although the application of this approach is still confounded by the techniques of oligonucleotide delivery into cells, currently most successfully used methods include loading of probes by the whole-cell patch-clamp technique and by microinjection. A transient permeabilization to load oligonucleotides into a cell population and to subsequently study responses at cellular level is currently tested. Such an approach would simplify the loading procedure, since loading by either sharp or patch pipettes is laborious. Nonetheless, already the antisense strategy contributed greatly to our understanding of stimulus-secretion coupling.

In the past it was thought that a pool of promiscuous GTP-binding proteins was involved in coupling several steps in signal transduction. However, with the discovery of multiple α, β, and γ-subunits of these proteins, it became possible that specific GTP-binding proteins were involved in specific pathways. A clear demonstration in support of this came with the introduction of the antisense strategy.[22,45,54] It was shown at cellular level that different surface receptors couple to VACCs via different heterotrimeric GTP-binding proteins. Moreover, subsequent steps in signal transduction such as the pathway leading from activation of TRH receptors to VACCs involves different GTP-binding proteins.[55] Future experiments should be directed to see whether different VACC α_1 subunits are modulated by different GTP-binding proteins. For this one should first determine the presence of multiple types of VACCs by the antisense strategy.

Recent progress in the identification of molecules interacting directly with the exocytotic machinery such as Rab3a, Rab3b and α_{i3} will undoubtedly bring us to the point where interactions between these molecules will have to be studied to understand their role in controlling exocytosis. How are these proteins coupled in mediating exocytosis?[68] These proteins are present in excitable and nonexcitable cells: is there a similarity in signaling leading to exocytosis mediated by GTP-binding proteins?

In conclusion, the application of antisense strategy did advance our understanding of stimulus-secretion coupling and we can expect in the future a growth of applications in vitro of this relatively new approach.

Acknowledgments

R.Z. is supported by a grant #J3 6027 from the Ministry of Sciences and Technology of the Republic of Slovenia. We would like to acknowledge Marjan Kadunc for photography, Sonja Grilc for cell cultures and Laura Kocmur-Bobanovic for comments on the manuscript.

REFERENCES

1. Douglas WW, Rubin RP. The role of calcium in the secretory response of adrenal medulla to acetylcholine. J Physiol 1961; 159:40-57.
2. Lledo P-M, Mason WT, Zorec R. Study of stimulus-secretion coupling in single cells using antisense oligodeoxynucleotides and patch-clamp techniques to inhibit specific protein expression. Cell Mol Neurobiol 1994; 14:539-556.
3. Crooke ST. Therapeutic applications of oligonucleotides. Annual Review of Pharmacology and Toxicology 1992; 32:329-376.
4. Baertschi AJ. Antisense oligonucleotide strategies in physiology. Molecular and Cellular Endocrinology 1994; 101 R15-R24.
5. Helene C, Toulme JJ. Specific regulation of gene expression by antisense, sense and antigene nucleic acids. Biochim Biophys Acta 1990; 1049: 99-125.
6. Crum C, Johnson JD, Nelson A, Roth D. Complementary oligodeoxynucleotide mediated inhibition of tobacco mosaic virus RNA translation in vitro. Nucleic Acids Research 1988; 16:4569-4581.
7. Ts'o PO. Nonionic oligonucleotide analogues (Matagen) as anticodic agents in duplex and triplex formation. Antisense Research & Development 1991; 1:273-276.
8. Morvan F, Rayner B, Imbach JL, Chang DK, Lown JW. alpha-DNA. I. Synthesis, characterisation by high field 1H-NMR, and base-pairing properties of the unnatural hexadeoxyribonucleotide alpha-[d(CpCpTpTpCpC)] with its complement beta-[d(GpGpApApGpG)]. Nucleic Acids Research 1986; 14:5019-5035.
9. van der Krol AR, Mol JN, Stuitje AR. Modulation of eukaryotic gene expression by complementary RNA or DNA sequences. Biotechniques 1988; 6:958-976.
10. Toulme JJ, Helene C. Antimessenger oligodeoxyribonucleotides: an alternative to antisense RNA for artificial regulation of gene expression—a review. Gene 1988; 72:51-58.
11. Dolnick BJ. Antisense agents in pharmacology. Biochemical Pharmacology 1990; 40:671-675.
12. Stein CA. Anti-sense oligodeoxynucleotides—promises and pitfalls. Leukemia 1992; 6:967-974.
13. Wahlestedt C. Antisense oligonucleotide strategies in neuropharmacology. Trends in Pharmacological Sciences 1994; 15:42-46.
14. Wagner E, Curiel D, Cotten M. Delivery of drugs, proteins and genes into cells using transferrin as a ligand for receptor-mediated endocytosis. Advanced Drug Delivery Reviews 1994; 14: 113-135.
15. Ts'O PO, Miller PS, Aurelian L, Murakami A, Agris C, Blake KR, Lin S-B, Lee BL, Smith CC. An approach to chemotherapy based on base sequence information and nucleic acid chemistry. Ann NY Acad Sci 1987; 507:220-241.
16. Stull RA, Taylor LA, Szoka Jr FC. Predicting antisense oligonucleotide inhibitory efficacy: a computational approach using histograms and thermodynamic indices. Nucleic Acids Research 1992; 20:3501-3508.

17. Marcus-Sekura CJ. Techniques for using antisense oligodeoxyribonucleotides to study gene expression. Analytical Biochemistry 1988; 172:289-295.
18. Eckstein F. Nucleoside phosphorothioates. Annual Review of Biochemistry 1985; 54:367-402.
19. Stein CA, Cheng YC. Antisense oligonucleotides as therapeutic agents—is the bullet really magical? Science 1993; 261:1004-1012.
20. Larrouy B, Blonski C, Boiziau C et al. RNase H-mediated inhibition of translation by antisense oligodeoxyribonucleotides: use of backbone modification to improve specificity. Gene 1992; 121:189-194.
21. Lledo P-M, Vernier P, Vincent J D, Mason WT, Zorec R. Inhibition of rab3B expression attenuates calcium dependent exocytosis in rat anterior pituitary cells. Nature 1993; 364:540-544
22. Kleuss C, Scherubl H, Hescheler J, Schultz G, Wittig B. Different β-subunits determine G-protein interactions with transmembrane receptors. Nature 1992; 358: 424-426.
23. Vlassov VV, Balakireva LA, Yakubov LA. Transport of oligonucleotides across natural and model membranes. Biochim Biophys Acta 1994; 1197:95-108.
24. Wagner RW. Gene inhibition using antisense oligonucleotides. Nature 1993; 372:333-335.
25. Heidenreich O, Kang S-H, Xu X, Nerenberg M. Application of antisense technology to therapeutics. Molecular Medicine Today 1995; 128-133.
26. Felgner PL, Gadek TR, Holm M, et al. Lipofection: a highly efficient, lipid-mediated DNA-transfection procedure. Proc Natl Acad Sci USA 1987; 84:7413-7417.
27. Iversen P. In vivo studies with phosphorothioate oligonucleotides: pharmacokinetics prologue. Anti-Cancer Drug Design 1991; 6:531-538.
28. Quattrone A, Di Pasquale G, Capaccioli S. Enhancing antisense oligonucleotide intracellular levels by means of cationic lipids as vectors. Biochemica 1995; 1:25-29.
29. Bennet CF, Chiang MC, Shoemaker JEE, Mirabelli K. Cationic lipids enhance cellular uptake and activity of phosphorotioate antisense oligonucleotides. Mol Pharmacol 1992; 41:1023-1033.
30. Wand S, Lee RJ, Cauchon G, Gorenstein DG, Low PS. Delivery of antisense oligodeoxyribonucleotides against the human epidermal growth factor receptor into cultured KB cells with liposomes conjugated to folate via polyethylene glycol. Proc Natl Acad Sci USA 1995; 92: 3318-3322.
31. Bhakdi S, Tranum-Jensen J, Sziegoleit A. Mechanism of membrane damage by streptolysin-O. Infection & Immunity 1985; 47:52-60.
32. Barry EL, Gesek FA, Friedman PA. Introduction of antisense oligonucleotides into cells by permeabilization with streptolysin O. Biotechniques 1993; 15:1016-1018.
33. Ahnert-Hilger G, Mach W, Fohr KJ, Gratzl M. Poration by (-toxin and streptolysin O: an approach to analyze intracellular processes. Methods Cell Biol 1989; 31:63-90.

34. Ruggiero M, Zimmerman TP, Lapetina EG. ATP depletion in human platelets caused by permeabilization with saponin does not prevent serotonin secretion induced by collagen. Biochem Biophys Res Com 1985; 131:620-627.
35. Koopmann Jr WR, Jackson RC. Calcium- and guanine-nucleotide-dependent exocytosis in permeabilized rat mast cells. Modulation by protein kinase C. Biochem J 1990; 265:365-373.
36. Knight DE, Baker PF. Calcium-dependence of catecholamine release from bovine adrenal medullary cells after exposure to intense electric fields. J Membrane Biol 1982; 68:107-140.
37. Tatham PE, Lindau M. ATP-induced pore formation in the plasma membrane of rat peritoneal mast cells. J Gen Physiol 1990; 95:459-176.
38. Lindau M, Gomperts BD. Techniques and concepts in exocytosis: focus on mast cells. Biochim Biophys Acta 1991; 1071:429-471
39. Saribas AS, Lustig KD, Zhang X, Weisman GA. Extracellular ATP reversibly increases the plasma membrane permeability of transformed mouse fibroblasts to large macromolecules. Analytical Biochemistry 1993; 209:45-52.
40. Hamill OP, Marty A, Neher E, Sakmann B, Sigworth FJ. Improved patch-clamp techniques for high-resolution current recording from cells and cell-free membrane patches. Pflugers Arch 1981; 391:85-100.
41. Marty A, Neher E. Tight-seal whole-cell recording. In: Sakmann B, Neher E, eds. Single Channel Recording. New York: Plenum Press, 1983:107-122.
42. Neher E, Marty A. Discrete changes of cell membrane capacitance observed under conditions of enhanced secretion in bovine adrenal chromaffin cells. Proc Natl Acad Sci USA 1982; 79:6712-6716.
43. Pusch M, Neher E. Rates of diffusional exchange between small cells and a measuring patch pipette. Pflugers Arch 1988;411:204-211.
44. Mobbs P, Becker D, Williamson R, Bate M, Warner A. Techniques for dye injection and cell labelling. In: Ogden D, ed. Microelectrode Techniques. The Plymouth Workshop Handbook. Cambridge: The Company of Biologists Ltd., 1994:361-389.
45. Kleuss C, Hescheler J, Ewel C, Rosenthal W, Schultz G, Wittig B. Assignment of G-protein subtypes to specific receptors inducing inhibition of calcium currents. Nature 1991; 353:43-48.
46. Leonetti JP, Mechti N, Degols G, Gagnor C, Lebleu B. Intracellular distribution of microinjected antisense oligonucleotides. Proc Natl Acad Sci USA 1991; 88:2702-2706.
47. Tatham PE, Gomperts BD. Late events in regulated exocytosis. Bioessays 1991; 13:397-401.
48. Penner R, Neher E. Secretory responses of rat peritoneal mast cells to high intracellular calcium. FEBS Letts 1988; 226:307-313.
49. Dolphin AC. Voltage-dependent calcium channels and their modulation by neurotransmitters and G proteins. Experimental Physiology 1995; 80:1-36.
50. Hepler JR, Gilman AG. G proteins. Trends in Biochemical Sciences 1992; 17:383-387.

51. Hille B. G protein-coupled mechanisms and nervous signalling. Neuron 1992; 9:187-195.
52. Neer EJ. Heterotrimeric G proteins: organizers of transmembrane signals. Neuron 1995; 80:249-257.
53. Baertschi AJ, Audigier Y, Lledo PM, Israel JM, Bockaert J, Vincent JD. Dialysis of lactotropes with antisense oligonucleotides assigns guanine nucleotide binding protein subtypes to their channel effectors. Molecular Endocrinology 1992; 6:2257-2265.
54. Kleuss C, Scherubl H, Hescheler J, Schultz G, Wittig B. Selectivity in signal transduction determined by γ subunits of heterotrimeric G proteins. Science 1993; 259:832-834.
55. Gollash M, Kleuss C, Hescheler J, Wittig B, Schultz G. Gi2 and protein kinase C are required for thyrotropin-releasing hormone-induced stimulation of voltage-dependent Ca^{2+} channels in rat pituitary GH3 cells. Proc Natl Acad Sci USA 1993; 90:6265-6269.
56. Campbell V, Berrow N, Dolphin AC. $GABA_B$ receptor modulation of Ca^{2+} currents in rat sensory neurones by the G protein Go: antisense oligonucleotide studies. J Physiol 1993; 470:1-11.
57. Berrow NS, Campbell V, Fitzgerald EM, Brickley K, Dolphin AC. Antisense depletion of β-subunits modulates the biophysical and pharmacological properties of neuronal calcium channels. J Physiol 1995; 482:481-491.
58. Perez-Reyes E, Schneider T. Calcium channels: structure, function, and classification. Drug Develop Res 1994; 33:295-318.
59. Tarroni P, Passafaro M, Pollo A, Popoli M, Clementi F, Sher E. Anti-β2 subunit antisense oligonucleotide modulate the surface expression of the (1 subunit of N-type (CTX sensitive Ca^{2+} channels in IMR 32 human neuroblastoma cells. Biochem Biophys Res Com 1994; 201:180-185.
60. Fournier F, Bourinet E, Nargeot J, Charnet P. Cyclic AMP-dependent regulation of P-type calcium channels expressed in Xenopus oocytes. Pflugers Arch 1993; 423:173-180.
61. Toutant M, Aunis D, Bockaert J, Homburger V, Rouot B. Presence of three pertussis toxin sensitive substrates and $G_o\alpha$ imunoreactivity in both plasma and granule membranes of chromaffin cells. FEBS Letts 1987; 215:333-344.
62. Yamamoto T, Furuki Y, Kebabian JW, Spatz M. α-Melanocyte-stimulating hormone secretion from permeabilized intermediate lobe cells of rat pituitary gland. FEBS Letts 1987; 219:326-330.
63. Nadin CY, Rogers J, Tomlinson S, Edwardson MJ. A specific interaction in vitro between pancreatic zymogen granules and plasma membranes: stimulation by G protein activators, but not Ca^{2+}. J Cell Biol 1989; 109:2801-2808.
64. Sikdar SK, Zorec R, Mason WT. Dual effects of G-protein activation on Ca^{2+}-dependent exocytosis in bovine lactotrophs. FEBS Letts 1989; 273:150-154.

65. Fischer von Mollard G, Südhoff TC, Jahn RA small GTP binding protein dissociates from synaptic vesicles during exocytosis. Nature 1991; 349:79-81.
66. Aridor M, Rajmilevich G, Beaven MA, Sagi-Eisenberg R. Activation of exocytosis by the heterotrimeric G protein G_{i3}. Science 1993; 262: 1569-1572.
67. Vitale N, Mukai H, Rouot B, Thierset D, Aunis D, Bader M-F. Exocytosis in Chromaffin cells: possible involvement of the heterotrimeric GTP-binding protein G_o. J Biol Chem 1993; 208:14715-14723.
68. Rupnik M, Zorec R. Intracellular Cl- modulates Ca^{2+}-induced exocytosis from rat melanotrophs through GTP-binding proteins. Pflugers Arch 1995; 431:76-83s.
69. Johannes L, Lledo PM, Roa M, Vincent JD, Henry JP, Darchen F. The GTPase Rab3a negatively controls calcium-dependent exocytosis in neuroendocrine cells. EMBO J 1994; 13:2029-2037.
70. Stettler O, Nothias F, Tavitian B, Vernier P. Double in situ hybridisation reveals overlapping neuronal populations expressing the low molecular weight GTPases Rab3a and Rab3b in rat brain. Eur J Neurosci 1995; 7:702-713.
71. De Matteis MA, Di Tullio G, Buccione R, Luini A. Characterization of calcium-triggered secretion in permeabilized rat basophilic leukemia cells. Possible role of vectorially acting G proteins. J Biol Chem 1991; 266:10452-10460.
72. Fewtrell C, Lagunoff D, Metzger H. Secretion from rat basophilic leukaemia cells induced by calcium ionophores. Effect of pH and metabolic inhibition. Biochim Biophys Acta 1981; 644:363-368.
73. Churcher Y, Gomperts BD. ATP-dependent and ATP-independent pathways of exocytosis revealed by interchanging glutamate and chloride as the major anion in permeabilized mast cells. Cell Regulation 1990; 1:337-346.
74. Rupnik M, Zorec R. Cytosolic chloride ions stimulate $Ca(2+)$-induced exocytosis in melanotrophs. FEBS Letts 1992; 303:221-223.
75. Wang HY, Watkins DC, Malbon CC. Antisense oligodeoxynucleotides to GS protein alpha-subunit sequence accelerate differentiation of fibroblasts to adipocytes. Nature 1992; 358:334-337.

CHAPTER 10

ANTISENSE TARGETING OF CORTICOTROPIN-RELEASING HORMONE AND CORTICOTROPIN-RELEASING HORMONE RECEPTOR TYPE I

Thomas Skutella, Joseph Christopher Probst,
Christian Behl and Florian Holsboer

The development of antisense strategies has generated considerable expectations in neurobiology (for reviews, see 1,2,3,4,5). Antisense application in the brain has become a new technology that might have a tremendous impact on the determination of molecular pathways and substrates of an organism's behavior that are controlled by independent stimuli. Designed originally for inhibition of oncogene expression, targeting of transcripts by dissecting their function with specific oligodeoxynucleotides (ODN) has proven to be a valuable tool in many cell biological approaches.

CRH AND ANXIETY

Anxiety is defined as a common emotional state that occurs during the body's first defense against harm. However, when anxiety becomes abnormally intense or prolonged it may cease to play a role in initiating adaptive responses and instead become a pathological condition. Anxiety disorders represent a major class of clinical problems that occur during the lifespan, often paving the way for abuse of substances

---

Antisense Strategies for the Study of Receptor Mechanisms,
edited by Robert B. Raffa and Frank Porreca. © 1996 R.G. Landes Company.

such as ethanol and benzodiazepines. Anxiety is a construct in which environmental factors and endogenous susceptibilities are interconnected in a complex form. The 41 amino acid neuropeptide corticotropin-releasing hormone (CRH), which is mainly produced in the hypothalamic paraventricular nucleus (PVN), is the key hormone for achieving hormonal stress adaptation through activation of the hypothalamic-pituitary-adrenal (HPA) system. Enhanced release of CRH from the median eminence (ME) into the hypophysial portal circulation is the major neuroendocrine response to a variety of stressors that result in increased HPA activity.[6] In pituitary corticotrophic cells, CRH induces synthesis and secretion of adrenocorticotrophic hormone (ACTH), β-endorphin and other proopiomelanocorticotropin (POMC)-derived peptides.[7,8,9,10,11] CRH neurons are not limited to the PVN and they have a widespread but selective distribution throughout the central nervous system (CNS).[12] High concentrations of CRH have been observed in the subcortical limbic system and brainstem and have been implicated in arousal and anxiety.[13] The anatomical distribution of CRH-expressing neurons in the rodent brain includes perikarya and projections in the amygdala, paraventricular nuclei, brainstem and spinal cord.[12] In the human brain, CRH immunoreactivity was found not only in the limbic system but also in the frontal, temporal and occipital cortex. These findings prompted a large number of studies supporting the hypothesis that CRH is also involved in neuroendocrine, behavioral and autonomic responses to stress (for reviews, see refs. 13,14). In particular, CRH was found to be anxiogenic in rats, and this effect was compensated by CRH antagonists or anxiolytic drugs such as benzodiazepines. This view is based on studies showing that intracerebroventricular (i.c.v.) administration of CRH produces anxiogenic effects in a novel open field,[15] an acoustic startle test,[16] a social interaction test,[17] and in paradigms of operant conflict and conditional emotional response.[18,19] Experimentally induced anxiogenic responses, e.g., ethanol withdrawal, can be suppressed by an i.c.v. administration of the CRH antagonist alpha-helical CRH-(9-41).[15] CRH antagonism in the amygdala suppressed stress-induced behaviors such as freezing.[20,21] Infusion of CRH into the locus coeruleus induced anxiety and related behaviors.[22] Benzodiazepines appear to exert at least part of their anxiolytic effects through suppressing CRH biosynthesis.[23] Because neuroendocrine function tests, sleep EEG studies and cerebrospinal fluid analyses all pointed to an overactivity of CRH neurons in human anxiety and depression, a CRH hyperdrive in conjunction with gradual vasopressin overexpression has been implicated in the pathogenesis of affective disorders (for review, see ref. 24). Many of the behavioral effects were observed after intracerebroventricular (i.c.v.) administration of pharmacological dosages of CRH and therefore the physiological significance of CRH remained unclear. Application of an antagonist yields only indirect information about the actual role of the respective receptor, because many antagonists tend to have binding affinities to related receptors as well. Furthermore, the efficacy of a competitive receptor antagonism also depends on the concentration of the endogenous ligand in the local extracellular fluid in the brain. Because CRH expression and release in diverse brain regions increase in response to stressful stimuli,[25] relatively high doses of antagonist are required to effectively reduce receptor-mediated effects of CRH. Alpha-helical CRH-(9-41) has been reported to produce dose-dependently either antagonistic, CRH-like agonistic or partially agonistic effects.[20,26,27]

Another tool to study central neuronal networks implicated in stress is an immuno-strategy. In vivo immunoneutralization of CRH in rats with an CRH antiserum or monoclonal antibodies biologically modulates effects of the HPA system.[28,29,30] Furthermore, combinations of CRH antibodies with nonlinked toxins have been injected above the PVN to manipulate ether stress-induced ACTH release.[31] Problems with this strategy include crossreactivity of the antisera and antibody used and immunostimulation of the host-animal after contact with the injected proteins. An alternative approach can be made to discriminate between the physiological involvement of a given neuropeptide and its receptors in behavioral function by using antisense ODNs, in which measurement of molecular, cellular and behavioral parameters can be conducted in the same organism.

SELECTION OF CRH ANTISENSE ODNS

Despite several advantages, phosphorothioate modification of ODN may reduce the selectivity of antisense targeting and also present the problem of toxicity and antigenicity in vivo. After administering complete phosphorothioated ODNs by the i.c.v. route into the brain, we observed changes reminiscent of sickness responses (e.g., curled body posture, piloerection, immobilization, reduced food and water intake) in most of the phosphorothioate antisense and sense treated animals and mixed bases controls.[32,33,34] These disturbing effects on the behavior of the animals were dose-dependent and hampered experimental designs. On the basis of such considerations, stable ODNs with minimum deviation from the native ODN structure have been introduced. These ODNs have 3'-3'-end inversion, show a half-life of 30 h in human serum and are capable of inhibiting protein expression in vitro and in vivo.[35]

In a first set of experiments,[36] an attempt was made to establish an animal model of reduced CRH activity in the CNS, using antisense DNA corresponding to the start coding region of rat CRH mRNA with either 3'-3'-inverted internucleotidic linkage or with all-phosphorothioate modification (Table 10.1). Animals bearing chronically implanted cannulas were injected i.c.v. with 50 μg ODN three times, 12 hours apart. As can be seen in Figure 10.1, after phosphorothioate sense ODN injections under basal conditions serum corticosterone levels

Table 10.1. Base composition of oligodeoxynucleotides (ODNs)

Oligonucleotide 5'-3' sequence	
CRH antisense studies	
rat CRH antisense	AGC CGC ATG GTT TAG GGG C
rat CRH mismatch	ATC CTC ATA GAT GAG GGG C
rat CRH antisense	AGC CGC ATG GTT TAG
rat AVP antisense	CAT GGC GAG CAT AGG TGG
rat OT antisense	CAC CAA CGC CAT GGC CTG
CRH receptor type I study	
rat antisense$_A$	CTG CGG GCG CCG TCC
rat sense$_A$	GGC CGG CGC CCG CAG
rat antisense$_B$	GCC GTC CCA TCC TCG GGC
rat mismatch $_A$	CTT CGG GTA ACG ACC
mouse antisense$_1$	CTG TCA GCG TCC TGG
mouse antisense$_2$	CAT CCT CGG GCT CGC
mouse sense$_2$	GCG AGC CCG AGG ATG
mouse mismatch $_2$	CAT AGT CGG GCA TAC

were significantly higher than vehicle (aCSF) or inversion-capped sense ODN controls. This increase was also apparent but less pronounced in phosphorothioate antisense-treated animals when compared with the corresponding sense group. After exposure to ether vapor, both phosphorothioate and inversion-capped antisense ODN-injected rats showed a significantly diminished stress-induced corticosterone secretion. These results indicated that (a) ether-stress induced corticosterone release is suppressed by i.c.v. CRH antisense treatment, (b) phosphorothioate ODNs might exert an unspecific, chronic stress-like activation of the HPA system and (c) this effect is partly inhibited by phosphorothioate antisense directed against CRH mRNA.

In another experiment[37] the interactive effects of CRH, arginine vasopressin (AVP) and oxytocin (OT) on the physiological regulation of the HPA system were investigated using antisense-inverted ODNs corresponding to rat CRH, AVP and OT. ODNs were infused bilaterally via an osmotic minipump system into the PVN over a 72 h period. The intracerebral infusions of the ODNs provided direct and restricted access to the target and avoided the dilution and rapid clearance associated with i.c.v. administration. The antisense ODNs reduced the respective neuropeptide content in the PVN, as revealed by immunoassay and no effects were observed on control peptides. Thirty minutes after exposure to ether vapor, CRH and combined CRH/AVP or CRH/OT antisense administration diminished stress-induced

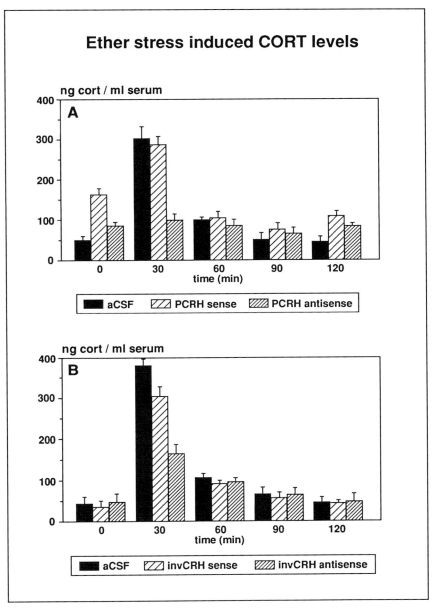

Fig. 10.1. Effect of CRH antisense, sense ODN and vehicle (aCSF) on serum corticosterone levels after ether vapor stress. Thirty min after exposure to stress phosphorothioate (a) and inversion-capped (b) antisense-injected rats showed markedly lower corticosterone levels than sense and vehicle injected rats. Diminishing stress-induced corticosterone plasma levels were similarly effective with phosphrothioate antisense and the 3' inversion-capped ODN. Ordinate: mg cortisol/ml plasma. Abscissa: timepoints of testing. Reprinted with permission from: Hormone Metabolic Research, 1994; 26:457-496.

corticosterone secretion compared to mismatch or vehicle-injected rats. AVP and OT alone and AVP/OT antisense only partially suppressed the ether vapor-induced corticosterone release. These results indicate that ether stress-induced corticosterone release is mediated mainly by CRH, mediated partially by AVP and OT and that AVP or OT antisense knock-outs do not potentiate the CRH antisense effect.

These data at least suggest the use of ODN analogs with minor deviations from the native oligonucleotide structure, and especially that complete phosphorothioate modifications introduce pharmacological consequences in vivo that do not depend on strict sequence complementarity to target mRNAs. To minimize side effects, it is reasonable to infuse i.c.v end-capped ODNs, because phosphorodiesters are rapidly degraded in the intraventricular compartment and we favor the use of unmodified ODNs to target intracerebral loci. In the following experiments, the modulation of neurochemical and behavioral components of the HPA system with CRH antisense ODNs is described.

CRH ANTISENSE ODNS AND ANXIETY

Although the administration of CRH produces anxiogenic effects, alpha-helical CRH-(9-41) produces no effects at lower doses and anxiogenic (rather than anxiolytic) effects at higher doses in the shuttle-box task.[16] Therefore, CRH antisense inversion-capped ODNs were tested in the shuttle-box procedure to determine whether behavioral changes correspond to changes in hypothalamic CRH expression.[38] The shuttle-box conflict task was used because it has been extensively validated for the assessment of anxiolytic and anxiogenic properties of various drugs.[39,40] Rats were equipped with a chronic implanted injection cannula and infused i.c.v. with inverted ODNs. Neuroendocrine effects were monitored by measuring CRH peptide levels as well as plasma ACTH and corticosterone. We observed an anxiolytic effect of the CRH antisense probe in the shuttle-box-avoidance procedure (Fig. 10.2A). CRH antisense-injected rats demonstrated accelerated acquisition of operant avoidance task and showed significantly more avoidance responses than sense and vehicle-treated animals. No statistically significant differences in locomotor activity, measured as numbers of intertrial crosses in the shuttle-box were observed between the experimental groups. Our data agree with several behavioral studies using anxiolytic drugs in the shuttle-box avoidance procedure.[20,39,40] These drugs improved performance during early acquisition of the discriminative avoidance response. The enhancement in avoidance performance has been attributed to reduction of freezing behavior, which occurs at the early stages of acquisition of this task.[41] Plasma ACTH and corticosterone levels were measured 60 min after exposure to the shuttle-box avoidance procedure to examine whether the antisense-induced change observed in the anxiety test was paralleled by changes in neuroendocrine parameters (Figs. 10.2B, C). Antisense-treated subjects showed significantly lower ACTH and

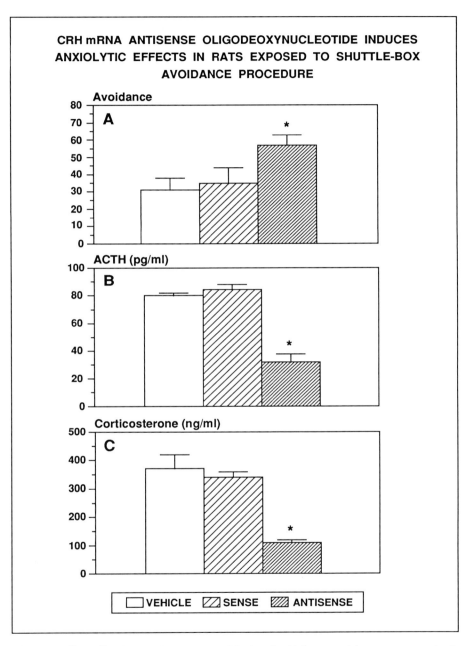

Fig. 10.2. Effect of invCRH antisense, sense ODN and vehicle on avoidance responses in the shuttle-box task (A). Antisense-injected animals showed significantly more avoidance responses than controls. ACTH (B) corticosterone (C) levels of rats directly after exposure to shuttle-box procedure. Antisense ODN-injected rats showed significantly lower ACTH and corticosterone levels than sense and vehicle injected rats. *$p < 0.001$ antisense vs. controls. Reprinted with permission from: Neuroreport, vol 5: 1627; 2183-2186.

corticosterone levels than did sense ODN and vehicle-treated animals. These stress-induced hormone levels reflect the intensity of aversive physical stimulation, correlated by the number of punished responses. Immunoreactive CRH content in homogenates of paraventricular nuclei was significantly reduced in antisense-treated animals, whereas controls did not show this effect. However, CRH amounts in the ME were similar in antisense-treated and control groups. Immunocytochemical staining for CRH was reduced in the PVN of antisense-treated rats, whereas immunostaining in the ME was apparently unchanged. Administration of the CRH antisense ODNs resulted in suppressed pituitary adrenocortical activity and reduced anxiety as evidenced by increased avoidance responses in the shuttle-box test. The shuttle-box challenge causes significant stress in rats resulting in a stimulation of the HPA system. Nevertheless, hypothalamic CRH pools proved to be far from totally depleted in this study. Although repeated injections of ODNs were used, we achieved only a partial reduction of hypothalamic CRH immunoreactivity. Although reduction of immunoreactive CRH levels in the PVN was about 50% there was no apparent loss of CRH in the ME, which is known to be the site of CRH release into the portal circulation. These findings, together with the significant reduction of plasma ACTH and corticosterone appear to indicate an interruption of secretory activity on antisense treatment, independent from the expected block of cytoplasmic CRH translation. The underlying mechanisms remain unclear.

In another set of experiments[42] we injected CRH antisense end-capped phosphorothioate ODNs and used social defeat stress to analyze the effect of CRH antisense ODNs in plus-maze performance combined with neuroendocrine parameters. The social confrontation of an unfamiliar intruder male rat into the home territory of an aggressive male resident normally results in a variety of defensive body postures and escape behaviors in the defeated intruder.[43] The fighting behavior that leads to the dominance of the resident activates the HPA system of the defeated animal which shows freezing behavior and reduced exploration.[44] At the behavioral level the anxiolytic effect of the CRH antisense could be demonstrated in the plus-maze task after social defeat (Fig. 10.3). The combination of social defeat and elevated plus-maze resulted in a significant increase in both time spent in the open arms and the entries made into open arms by antisense-treated animals than by vehicle- and sense-injected rats. Antisense-treated subjects spent more time and made more entries into the open arms, indicating a reduced anxiety level. It must be emphasized that the antisense probe did not alter plus-maze performance of unstressed rats. These observations suggest that the effect of CRH antisense ODNs may be stress-dependent or more evident when this particular system is challenged. This finding may be related to a general mechanism of central neuropeptidergic hormones, that are modulatory rather than activating.

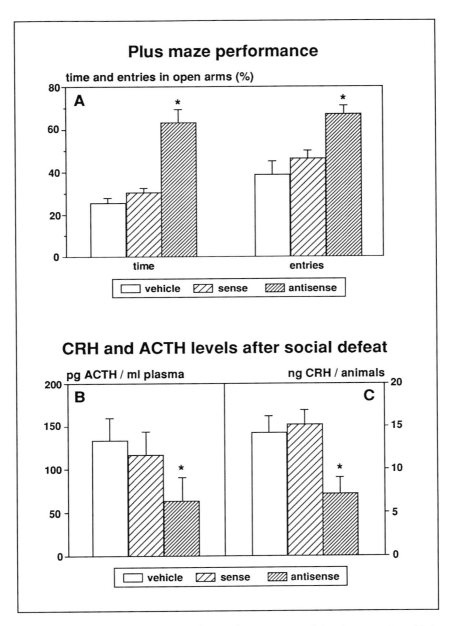

Fig. 10.3. Time spent and entries made into the open arms of the plus maze for vehicle, sense and antisense oligonucleotide treated animals, which had been previously exposed to an aggressive male resident for 10 min in the social defeat paradigm (A) *p < 0.01. Solid-phase immunoassay (SOFIA) for CRH content (ng CRH/animals) in homogenates of microdissected PVNs (B) In CRH-antisense treated rats lower amounts of CRH was observable than in sense and vehicle controls. * p < 0.001. ACTH plasma levels 30 min after exposure to social defeat (C) Antisense ODN injected rats showed significantly lower ACTH levels compared to sense and vehicle injected rats. *p < 0.001 antisense vs. controls. Reproduced with permission from: Cellular and Molecular Neurobiology 1994; 14: 5, 579-88.

Furthermore, it must be noted that the effect of the CRH antisense probe was not accompanied by sedation because we observed consistent overall activity. It has been shown that social conflict activates the HPA system as well as central stress response and related behaviors in rats.[45] The elevated plus-maze is used to assess anxiogenic consequences of natural stressors, such as social defeat, that reduces untreated rat's exploration behavior of open arms of the plus-maze allows them to establish a preference for enclosed maze compartments.[46] The observed reduction of anxiety among social-defeated rats through suppression of CRH biosynthesis by CRH antisense ODNs agrees with results obtained with CRH antagonists and benzodiazepine anxiolytics. Britton et al[18,47] have shown that the i.c.v. administration of CRH to rats enhances behavioral effects of novelty and that chlordiazepoxide attenuates CRH-induced response suppression in a conflict test. Moreover, Butler et al[22] documented that CRH produces fear-enhancing effects in a dose dependent fashion after infusion into the locus coeruleus of rats. This effect was suppressed by administration of anxiolytic drugs such as alprazolam or adenazolam.[23] A study by Kalin et al[20] showed that antagonizing the stress-induced activation of CRH neurosecretion reduces the stress-induced freezing behavior of rats.

Our study indicates that the modulation of the HPA system with CRH antisense ODN affects neuroendocrinological components of the stress response as well as central circuits that are thought to mediate respective behaviors.[12] The major finding of the behavioral studies in CRH antisense-treated rats is the reduction of anxiety during shuttlebox performance and after social defeat. It can be assumed that the i.c.v.-infused antisense ODNs readily crossed neuronal cell membranes in paraventricular regions.[48,49] The antisense treatment apparently affected only a part of the total peptide content, decreasing newly synthesized and processed CRH. Nevertheless, our results suggest that stress-induced release of ACTH was abolished because of a lack of CRH in portal blood. Consequently, the antisense ODN may abolish the enhanced stress-induced release of ACTH.

These findings indicate that in vivo antisense targeting of CRH mRNA effectively blocked the HPA system thus providing a valuable strategy for the manipulation of central and systemic stress response. Against this background the specific block of CRH activity achieved by ODN administration raises the question of its therapeutic potential. Several issues ranging from route of administration, protection from degrading enzymes and costs of production need to be addressed. If these problems can be solved, it will be easy to envisage that antisense strategies will be exploited not only for physiological studies. It may also open up a new lead in pharmaceutical drug design based on the structure of genes. Perhaps the most interesting insights into pathogenesis of psychiatric disorders has been obtained from studies on CRH physiology in animals and humans.[50] Collectively these studies conclude that the HPA system is profoundly disturbed in depression and anxiety disorders. Moreover, clinical and preclinical studies have underscored that antidepressants probably act through rectifying excessive HPA activity driven by over- production of CRH.[51,52] Because CRH is a prime candidate as mediator of psychopathology and neuroendocrinological symptoms of affective disorders, strategies targeted against biosynthesis of CRH and its receptor appear to have the strongest potential to provide a new class of antidepressant drugs.

CRH RECEPTOR (TYPE I) AND ANXIETY

CRH initiates its biological effects by binding to membrane receptors that are widely distributed throughout the nervous system, as well as in the pituitary, adrenal glands and spleen as shown by autoradiographic mapping.[53,54,55] CRH-elicited endocrinological and behavioral responses to stressful environments can be attenuated or reversed by administration of CRH receptor antagonists.[14,20,26,56,57] Two rat brain CRH receptor (CRH-R) cDNAs have been cloned (CRH-R type I[58] and CRH-R type II[59]). In situ hybridizations documented a prominent CRH-R type I mRNA expression in the cerebellar- and neocortex, olfactory bulb, hypothalamus and anterior pituitary, but only minor expression in the amygdala.[60,61] CRH-R type II mRNA is highly expressed in the ventromedial nuclei of the hypothalamus, the lateral septum, the amygdala, and the entorhinal cortex. No CRH-R type II expression is detectable in the cerebellar- and neocortex and in the pituitary lobes.[59] The obvious differences in tissue distribution between CRH-R type I and CRH-R type II may predict important basic physiological differences between these two receptor subtypes. The CRH-R type I is a member of a family of transmembrane receptors that comprise receptors for calcitonin, vasoactive intestinal peptide and parathyroid hormone.[62] The CRH-R type I is coupled to a GTP-binding protein mediating the increase of intracellular cAMP after its binding of CRH, its natural ligand. This rise in second messenger cAMP level can be inhibited by the CRH antagonist alpha-helical CRH-(9-41). At the behavioral level it has been shown that an i.c.v. administration of CRH dose-dependently produces marked behavioral inhibition in novel stressful environments, such as elevated plus-maze, where CRH-R antagonist alpha-helical CRH-(9-41) produces opposite effects.[14] However it is not yet known which CRH-R subtype mediates the different behavioral actions in vivo. Also the complex pharmacological profile of CRH-Rs is not completely elucidated.

To select potent antisense ODNs the efficacy of different antisenses studied in rat primary pituitary cell culture and in clonal mouse pituitary cells (ATt-20). To detect different functions of the CRH-R subtypes I and II in vivo, we used end-capped phosphorothioate antisense ODNs targeted specifically against the CRH-R type I mRNA. Uptake of different sequences of end-capped phosphorothioate antisense corresponding to CRH-R type I mRNA and mismatch ODNs (Table 10.1) were examined in rat anterior primary pituitary cell culture. After addition of 5-10 µM FITC-labeled mismatch and antisense ODNs to the cells we observed intracellular accumulation of ODNs in the cells after 4 to 24 hours, indicating an effective loading of cells with ODNs. To investigate possible toxic effects of intracellularly accumulated ODNs, dead cells were labeled by incubation with propidium iodide (PI). This showed that most of the FITC-positive ODN-containing cells were PI-negative and therefore still alive. Intracellular uptake of ODN concentrations as low as 1 µM were increased by a cationic transfection system (Fig. 10.4). The mechanism by which ODN traverses the plasma membrane is still unclear, but the observations made so far suggest a caveolar, proteolytic transport rather than endocytosis.[63] An electron microscopic study of ODN transport across eukaryotic cell

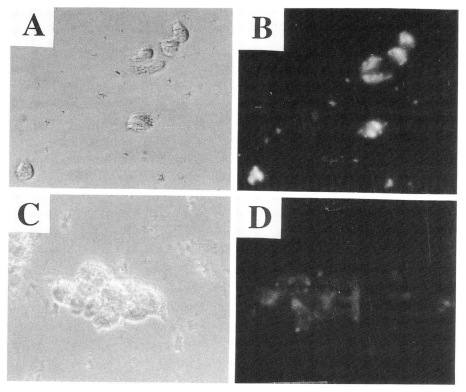

Fig. 10.4. Uptake of antisense ODN as detected by fluorescence microscopy. Primary pituitary cell culture was incubated with a mixture of 10 mM DOTAP and 1 µM of FITC-labeled antisense ODN or 1 µM of FITC-labeled antisense ODN alone. After 24 h microscopy using green filters (FITC) was performed. A-B show corresponding optical fields of antisense/DOTAP, C-D corresponding antisense: A/C phase contrast, C/D FITC.

membranes on the cell culture level[64] has revealed that ODNs cross the external cell membrane, traverse the cytosol and begin to enter the cell nucleus. After 30 to 60 minutes of incubation with ODNs staining was observed inside the nuclear membrane as well as throughout the nucleus associated with euchromatin. No sequestration of labeled ODNs could be found in cytoplasmic vesicles en route from exterior of the cell to the nucleus. Comparison of cellular uptake and binding of FITC-conjugated ODN showed that phosphorothioate ODN had the highest cell binding and uptake capacity followed by end-capped phosphorothioate ODNs, phosphodiesters and methylphosphonate-phosphodiester (unpublished results). Confocal microscopy confirmed intracellular uptake of ODN and indicated little nuclear uptake by 4 hours.

After 24, 48 and 72 hours of repeated ODN treatment, cell survival was not significantly decreased as assessed with MTT (less than 10%). Addition of a transfection system (10 μM) with and without 1 μM ODN did also not affect cell survival. As there was only minor toxicity of applied ODNs in vitro, we concluded that the selected antisense are useful tools for further studies on effectiveness of translational inhibition in vitro and in vivo. As revealed by RT-PCR, repeated application of CRH-R type I antisense probes over a 72 h period to primary anterior pituitary cell culture produced sequence- and dose-dependent increases in CRH-R type 1 mRNA. CRH-R binding was significantly reduced in rat primary pituitary cells after a 48 h treatment with 10 μM antisense$_A$. After 72 of antisense$_A$ treatment, no further reduction in CRH-R binding was obtained. Effective antisense$_A$ concentrations could be lowered again to 1 μM when ODNs were used in combination with the transfection system. To further investigate reductions in the physiological function of CRH-R type I with specific receptor antisense ODNs, CRH-stimulated ACTH-release was studied. Repeated exposure of primary pituitary cells to 10 μM antisense$_A$ ODNs for 72 h period attenuated CRH stimulated ACTH-release. When antisense$_A$ was mixed with the sequence complementary sense$_A$ prior to the addition to cells, this antisense effect was completely abolished, indicating this effect's specificity. Similar effects on CRH binding and ACTH release in primary pituitary culture were observed by Owens et al[65] with an phosphodiester ODN d(CTG CGG GCG GCG TCC), utilizing the same effective sequence as selected for rat CRH-R type I cDNA in the study described here.

Distribution of i.c.v. infused antisense$_A$ ODNs over 72 h (0.5 μg/0.5μl/h) into the rat brain was most pronounced around the injection site and the ventricular compartments, such as the paraventricular region, as assessed by in situ hybridization with sense probe.[7] Antisense ODNs could not be detected in the pituitary lobes. Antisense ODN administered over a 72 h period bilaterally into the amggdala showed distribution of ODNs around the injection sites, but ODNs also diffused along the injection cannula, diffused into the ventricular compartment and partially into the periventricular brain parenchyma. Therefore behavioral effects observed might not be related purely to the targeted anatomical location.

To achieve stable and effective concentrations of the ODNs at the site of intended action in the in vivo experiments, endonuclease stable endcapped phosphorothioate-modified antisense$_A$ ODNs were infused over 72 h period by osmotic minipump system. The time schedule and mode of administration was chosen to allow for effective hybridization of the antisense ODN with the target mRNA. This permitted the loss of preexisting functional CRH-R type I as part of the general turnover of membrane proteins and also counteraffected possible compensatory regulations, as indicated by the increased new synthesis of CRH-R type I mRNA both in vitro and in vivo. Though the precise mode of action of the antisense ODNs is not completely known, our data suggest that antisense effects are primarily based on the induction of translational arrest, indicated by the reduced CRH ligand binding and ACTH release in vitro. Also the antisense effects in vivo resulted in reduced CRH ligand binding and produced behavioral effects that were associated with increased CRH-R mRNA levels, indicating ongoing transcriptional activity. Although the reduction in CRH-binding as well as ACTH-release —the physiological response of CRH-R activation— is significant, our antisense approach is not a total "knockout." The experimental results reconfirm data from other in vitro and in vivo antisense receptor experiments, demonstrating that after antisense treatment receptor pools were not totally depleted.[66,67,68,69,70] Because CRH has been suggested to be physiologically involved in the regulation of avoidance and escape behaviors (for reviews see refs. 13,14,71), we examined consequences of CRH-R antisense infusions in anxiety models after i.c.v. CRH administration. It has been demonstrated that CRH-treated rats do not enter the center of a novel open field apparatus.[72] Similar results were obtained with food-deprived rats in a novel open field with food placed into the central portion.[73] Furthermore, it has been demonstrated in rats that CRH dose-dependently reduced the time spent in the open arms and entries made into in the open arms of the elevated plus-maze thus displaying a general anxiogenic-like profile.[72] The CRH antagonist alpha-helical CRH-(9-41) produces opposite effects.[13] Antisense$_A$ infusion over 72 h significantly attenuated anxiogenic behavior normally induced after injection of CRH in behavioral test situations such as the elevated plus-maze and the novel open field test. After i.c.v. administration of CRH (100 pM) control animals showed a preference for closed compartments of the elevated plus maze, demonstrating usage of the anxiety paradigm. In contrast, antisense$_A$-treated animals made significantly more entries into the open arms and spent more time in the open arms than controls (vehicle, mismatch$_A$, antisense$_B$), which suggests an anxiolytic effect of this treatment. In addition, this anxiolytic effect of the CRH-R antisense$_A$ probe could be confirmed in the novel open field test after i.c.v. CRH application. Antisense$_A$ treated rats made more entries into the inner section of the open field and showed higher levels of locomotor activity. Control animals moved only hesitantly in the outer sections. The results clearly indicate that the modulation of this CRH-R subtype modifies the impact of CRH on the behavior of the animal.

Because several behavioral effects induced by CRH were demonstrated to be independent of an activation of the HPA system,[74,75,76] an involvement of extrahypothalamic CRH receptors in other limbic regions has been postulated. An important neuronal center in pathways underlying integration of anxiogenic behaviors is the central nucleus of the amygdala (CeN).[77,78] The CeN contains CRH positive cells and fibers in anatomical connections to the neocortex, hippocampus, hypothalamus and nuclei in the brain stem. Swiergiel et al[21] and Rassnik et al[79] blocked stress induced behavior by antagonism of CRHRs in the central amygdala of the rat. Heinrichs et al[26] could attenuate the anxiogenic effect of an intruder rat socially confronted with an aggressive male resident in its home territory by local infusion of alpha-helical CRH-(9-41) into the CeN. This result obtained with antagonist treatment could be confirmed by Liebsch et al[80] with antisense ODNs targeting CRH receptor type 1 in the CeN. CRH receptor type I antisense ODNs d(GCC GTC CCA TCC TCG GGC) that were infused bilaterally directly into the CeN over 4 days reduced social defeat-induced anxiety in rats in the elevated plus-maze.

The behavioral data support first the hypothesis that the CRH-R is a major player in the central regulation of anxiety and second demonstrate the specific importance of receptor subtype I. Whether other receptor-subtypes take part in anxiety regulation remains unknown. Therefore, it appears that an antisense approach in general represents a valuable molecular tool to dissect the putative varied roles of the different receptor subtypes and may also help in the design of specific receptor antagonists. Our data stress the importance of a drug design that specifically downregulate CRH and CRH-R type I function as anxiolytics.

CONCLUSION

Antisense targeting may be a useful tool in behavioral neuroscience that supplements conventional techniques by inducing specific deficits on the peptide level. To establish efficient antisense models, it is our opinion that testing selected ODNs for potency and specificity in simplistic environments, such as cell culture systems, is absolutely required. The selected ODNs may then be extrapolated to more complex tasks such as in vivo models. Reversible molecular deficits could be established and topographically analyzed by applying antisense ODNs in combination with other techniques in special target regions. This may lead to new insights in the control of neuronal functions that include diverse respondent and operant behaviors.

REFERENCES

1. Goodchield J. Inhibition of gene expression by oligonucleotides. In: Cohen JS, ed. Oligonucleotides Antisense Inhibitors of Gene Expression. Houndsmills: Macmillan Press, 1989:53-77.
2. Stein CA, Cheng Y-C. Antisense oligonucleotides as therapeutic agents—is the bullet really magical? Science 1993; 261:1004-1012.
3. McCarthy MM. Use of antisense oligodeoxynucleotides to block gene expression in the central nervous system. In: de Kloet ER and Sutanto W, eds., Neurobiology of Steroid, Methods in Neuroscience 1994:342-358.
4. Wagner R. Gene inhibition using antisense oligodeoxynucleotides. Nature 1994; 372:333-335.
5. Morris M and Lucion AB, Antisense Oligonucleotides in the study of neuroendocrine systems. J. Neuroendocrinology 1995; 7:403-500.
6. Imaki T, Vale W. Chlordiazepoxide attenuates stress-induced accumulation of corticotropin-releasing factor mRNA in the paraventricular nucleus. Brain Res 1993; 623:223-228.
7. Gibbs DM, Steward RD, Vale W, Rivier J, Yen SSC, Synthetic corticotropin-releasing factor stimulates secretion of immunoreactive beta-endorphin/beta-lipoprotein and ACTH by human fetal pituitaries in vitro. Life Sci. 1982; 32:547-550.
8. Rivier C, Brownstein J, Spiess J, Rivier J, Vale WW. In vivo corticotropin-releasing factor-induced secretion of adrenocorticotropin, beta-endorphin, and corticosterone. Endocrinology 1982; 110:272-278.
9. Meunier H, Lefevre G, Dumont D, Labrie F. CRF stimulates alpha-MSH secretion and cyclic AMP accumulation in pars intermedia cells. Life Sci 1982; 31:2129-2135.
10. Proulx-Ferland L, Labrie F, Dumont D, Cote J, Coy DH, Sveiraf J. Corticotropin-releasing factor stimulates secretion of melanocyte-stimulating hormone from the rat pituitary. Science 1992; 217:62-63.
11. Grossman A, Kruseman ACN, Perry L, Tomlin S, Schally AV, Coy DH, Rees LH, Comaru-Schally AM, Besser GM. New hypothalamic hormone, corticotropin-releasing factor, specifically stimulates the release of adrenocorticotropic hormone and cortisol in man. Lancet 1982; 1:921-922.
12. Swanson LW, Sawchenko PE, Rivier J, Vale WW. Organization of ovine corticotropin-releasing factor immunoreactive cells and fibers in the rat brain: an immunohistochemical study. Neuroendocrinology 1983; 36:165-186.
13. Dunn AJ, Berridge CW. Physiological and behavioral response to corticotropin-releasing factor administration: is CRF a mediator of anxiety or stress response? Brain Res Rev 1990; 15:71-100.
14. Owens MJ, Nemeroff CB. The physiology and pharmacology of corticotropin-releasing factor. Pharmacol Rev 1991; 43:425-473.
15. Baldwin HA, Rassnik S, Rivier J, Koob GF, Britton KT. CRF antagonist

reverses the anxiogenic response to ethanol withdrawal in the rat. Psychopharmacology 1991; 103:227-232.
16. Swerdlow NR, Geyer MA, Vale W, Koob GF. Corticotropin-releasing factor potentiates acoutic startle in rats: blockade by chlordiazepoxide. Psychopharmacology 1986; 88:147-152.
17. Dunn AJ, File SE. CRF has an anxiogenic action in the social interaction test. Horm Behav 1987; 21:193-202.
18. Britton DR, Koob GF, Rivier J, Vale W. Intraventricular corticotropin-releasing factor enhances behavioral effects of novelty. Life Sci 1982; 31:363-367.
19. Cole BJ, Koob GF. Propanol antagonizes the enhanced conditioned fear produced by corticotropin-releasing factor. J Pharmacol Exp Ther 1988; 47:902-910.
20. Kalin NH, Sherman JE, Takahashi LK. Antagonism of endogenous CRH systems attenuates stress induced freezing behavior in rats. Brain Res 1988; 457:130-135.
21. Swiergiel AH, Takahashi LK, Kalin N. Attenuation of stressed-induced behavior by antagonism of corticotropin-releasing factor receptors in the central amygdala of the rat. Brain Res 1986; 623:229-234.
22. Butler PD, Weis JM, Shout JC, Nemeroff CB. Corticotropin-releasing factor produces fear-enhancing and behavioral activating effects following infusion into the locus coeruleus. J Neurosci 1990; 10:176-183.
23. Owens MJ, Bissette G, Nemeroff BB. Acute effects of alprazolam and adinazolam on the concentrations of corticotropin-releasing factor in the rat brain. Synapse 1989; 4:196-202.
24. Holsboer F, Spengler D, Heuser I. The role of corticotropin releasing hormone in the pathogenesis of Cushing's disease, anorexia nervosa, alcoholism, affective disorders and dementia. Prog Brain Res 1992; 93:385-417.
25. Kalin NH, Takahashi LK, Chen FL. Restraint stress increases corticotropin-releasing hormone mRNA content in the amygdala and paraventricular nucleus. Brain Res 1994; 656:182-186.
26. Heinrichs SC, Pich EM, Miczek KA, Britton KT, Koob GF. Corticotropin-releasing factor antagonist reduces emotionality in socially defeated rats via direct neurotropic action. Brain Res.1992; 581:190-197.
27. Menzaghi F, Howard RL, Heinrichs SC, Vale W, Rivier J, Koob GF. Characterization of a novel and potent corticotropin-releasing antagonist in rats. J Pharmacol Exp Ther 1994; 269:564-572.
28. Rivier C, Rivier J, Vale W. Inhibition of adrenocorticotropin hormone secretion in rat by immunoneutralisation of corticotropin-releasing factor. Science 1982; 218:377.
29. Linton EA, Tilders FJH, Hodgkinson S, Berkenbosch F, Vermes I, Lowry PJ. Stress induced secretion of adrenocorticotropin in rats is inhibited by administration of antisera to ovine corticotropin-releasing factor and vasopressin. Endocrinology 1985; 116:966-970.
30. Burlet AJ, Menzaghi F, Tilders FJ, van Oers JW, Nicolas JP, Burlet CR.

Uptake of a monoclonal antibody to corticotropin-releasing factor (CRF) into rat hypothalamic neurons. Brain Res 1990; 517:283-293.

31. Menzaghi F, Burlet A, Van Oers JWAM, Barnanel G, Tilders FJH, Nicolas JP, Burlet C. A new perspective for the study of central neuronal networks implicated in stress. In: Kvetnansky R, McCarthy R, Axelrod J, eds., Neuroendocrine and Molecular Approaches. Stress Gordon and Breach Science. 1992.

32. Skutella T, Probst JC, Jirikowski GF, Holsboer F, Spanagel R. Ventral tegmental area (VTA) injections of tyrosine hydroxylase phosphorothioate antisense oligonucleotide suppress operant behavior in rats. Neurosci Lett1994a; 167:55-58.

33. Skutella T, Probst JC, Engelmann M, Wotjak CT, Landgraf R, Jirikowski GF. Vasopressin antisense oligonucleotide induces temporary diabetes insipidus in rats. J Neuroendocrinology 1994b; 6:121-125.

34. Schöbitz B, Engelmann M, Pezeshki G, Landgraf R, Montkowski A, Probst JC, Skutella T, Spanagel R, Stöhr T, Wotjak C, Reul JMHM, Holsboer F. Centrally administered oligodeoxynucleotides in rats: influence of their nonspecific effects on behavioral studies. J Pharmacol Exp Ther, in press.

35. Ortigao JFR, Rosch H, Selter H, Frohlich A, Lorenz A, Montenarh M, Seliger H. Antisense effect of oligonucleotides with inverted terminal internucleotidic linkages: a minimal modification protecting against nucleolytic degradation. Antisense Res Dev 1992; 2:129-146.

36. Skutella T, Stöhr T, Probst JC, Ramalho-Ortigao FI, Holsboer F, Jirikowski GF. Antisense oligodeoxynucleotides for in vivo targeting of corticotropin-releasing hormone mRNA: comparison of phosphorothioate and 3' inverted probe performance. Horm Metab Res 1994c; 26:460-464.

37. Skutella T, Probst JC, Holsboer F, Jirikowski GF. Effects of intracerebral CRH, AVP and OT antisense infusions on corticosterone secretion after ether stress in the rat. Submitted for publication.

38. Skutella T, Probst JC, Criswell H, Moy C, Breese G, Jirikowski GF, Holsboer F. Corticotropin-releasing hormone antisense oligodeoxynucleotide induce induce anxiolytic effect in rats. Neuro Report 1994d; 5:2181-2185.

39. Thompson RC, Seasholtz AF, Herbert E. Rat corticotropin-releasing hormone gene: sequence and tissue-specific expression. Mol Neuroendocrinology 1987; 1:363-370.

40. Robichaud RC, Sledge KM, Hefner MA, Goldberg ME. The influence of SKF-525-A on the acute pharmacological properties of chlordiazepoxide. Psychopharmacoligia 1973; 32:157-160.

41. Fernandez-Teruel A, Escorihuela RM, Nunez JF, Zapata A, Boix F, Salazar W, Tobena A. Brain Res Bul 1991; 26:173-176.

42. Skutella T, Montkowski A, Stöhr T, Probst JC, Landgraf R, Holsboer F, Jirikowski GF. Corticotropin-releasing hormone (CRH) antisense oligodeoxynucleotide treatment attenuates social defeat-induced anxiety in rats. Cell Mol Neurobiol 1994e; 14:579-588.

43. Miczek KA. A new test for aggression in rats without aversive stimula-

tion: differential effects of d-ampethamine and cocaine. Psychopharmacology 1979; 60:253-259.
44. Shuurmann T. Hormonal correlates of agonistic behavior in adult male rats. In: McConnell PS, Boer GJ, Romijn HJ, van de Poll NE, Corner MA, eds. Progress in Brain Research 1986; 53, Amsterdam: Elsevier, 1986:53.
45. Huhman KL, Bunnell BN, Mougey EH, Meyerhoff JL. Effects of social conflict on POMC-derived peptides and glucocorticoids in male goldhamsters. Physiol Behav 1990; 47:949.
46. Pellow S, Chopin P, File SE, Briley M. Validation of open: closed arm entries in the elevated plus maze as a measure of anxiety in the rat. J Neurosci Methods 1985; 14:149-167.
47. Britton KT, Morgan J, Rivier J, Vale W, Koob GF. Chlordiazepoxide attenuates CRF-induced responses suppression in conflict test. Psychopharmacology 1985; 86:170-174.
48. Whitesell L, Geselowitz D, Chavany C, Fahmy B, Walbridge S. Stability, clearance, and disposition of intraventricularly administered oligodeoxynucleotides: implications for therapeutic application within the central nervous system. Proc Natl Acad Sci 1993; 90:4665-4669.
49. Yee F, Ericson H, Reis DJ, Wahlestedt C. Cellular uptake of intracerebroventricularly administered biotin- or digoxigenin-labeled antisense oligonucleotides in the rat. Cel Mol Neurobiol 1994; 14:475-486.
50. Kling MA, Rubinow DR, Doran AR, Roy A, Davis CL, Calabrese JR, Nieman LK, Post RM, Chrousos GP, Gold PW. Cerebrospinal fluid immunoreactive somatostatin concentrations in patients with Cushing's disease and major depression: relationship to corticotropin-releasing hormone and cortisol secretion. Neuroendocrinology 1994:57, 79-88.
51. Pepin MC, Pothier F, Barden N. Antidepressant drug action in a transgenic mice model of the endocrine changes seen in depression. Mol Pharmacol 1992; 42:991-995.
52. Reul JMHM, Stec I, Söder M, Holsboer F. Chronic treatment of rats with the antidepressant amitriptyline attenuates the activity of the hypothalamic-pituitary-adrenocortical system. Endocrinology 1993; 133: 312-320.
53. DeSouza EB, Perrin MH, Rivier JH, Vale WW, Kuhar MJ. Corticotropin-releasing factor receptors in rat pituitary gland: autoradiographic localization. Brain Res 1984a; 296:202-207.
54. DeSouza EB, Perrin MH, Insel TR, Rivier J, Vale WW, Kuhar MJ. Corticotropin-releasing factor receptors in rat forebrain: autoradiographic identification. Science 1984b; 224:1449-1451.
55. DeSouza EB, Insel TR, Perrin MH, Rivier J, Vale WW, Kuhar MJ. Corticotropin-releasing factor receptors are widely distributed within the rat central nervous system. An autoradiographic study. J Neuroscience 1984c; 5:3189-3203.
56. Koob JF, Heinrichs SC, Pich EM, Menzaghi F, Balswin H, Miczek K,

Britton KT. Corticotropin-releasing factor. In: Ciba Foundation Symposium. Chichester: Wiley, 1993:277-295.
57. Takahashi LK, Kalin NH, Vanden Burg JA, Sherman JE. Corticotropin-releasing factor modulates defensive withdrawal and exploratory behavior in rats. Behav Neurosci 1989; 103:648-654.
58. Perrin MH, Donaldson CJ, Chen R, Lewis KA, Vale WW. Cloning and functional expression of a rat corticotropin-releasing factor (CRF) receptor. Endocrinology 1986; 133:3058-3061.
59. Lovenberg TW, Liaw CW, Grigoriadis DE, Clevenger W, Chalmers DT, DeSouza EB, Oltersdorf T. Cloning and characterization of a functionally distinct corticotropin-releasing factor receptor subtype from rat brain. Proc Natl Acad Sci USA 1995; 92:836-840.
60. Potter E, Sutton S, Chen R, Perrin M, Lewis K, Sawchenko PE, Vale W. Distribution of corticotropin-releasing factor receptor mRNA expression in the rat brain and pituitary. Proc Natl Acad Sci USA 1994; 91:8777-8781.
61. Wong M-L, Licinio J, Paternak KI, Gold PW. Localization of corticotropin-releasing hormone (CRH) receptor mRNA in adult rat brain by in situ hybridization histochemistry. Endocrinology 1994; 135:2275-2278.
62. Bilezikjian LM and Vale WW, Glucocorticoids inhibit corticotropin-releasing factor- induced production of adenosine 3', 5'-monophosphate in cultured anterior pituitary cells. 1983; 113:657-662.
63. Beltinger C, Saragovi HU, Smith RM, LeSauteur L, Shah N, DeDionisoi L, Christensen L, Raible A, Jarett L, Gewirtz AM. Binding, uptake, and intracellular trafficking of phosphorothioate-modified oligonucleotides. J Clin Invest 1995; 95:1814-1823.
64. Temsamani J, Metelev V, Levina A, Agraval S, Zamecnik P. Inhibition of in vitro transcription by oligonucleotides. Antisense Res Dev 1994; 4:279-284.
65. Owens MJ, Mulchahey JJ, Kasckow JW, Plotzky PM, Nemeroff CB. Exposure to an antisense oligonucleotide decreases corticotropin-releasing factor binding in rat pituitary cultures. J Neurochem 1995:5:2358.
66. Skutella T, Stöhr T Probst JC, Heuser I, Jirikowski GF. Modulation of the hypothalamic-pituitary-adrenocortical (HPA) system by antisense oligodeoxynucleotides to corticotropin-releasing hormone mRNA. 19th CINP Congress, Washington DC, 1994f: abstract.
67. Skutella T, Probst JC, Caldwell J, Pederson CA, Jirikowski GF. Antisense oligodeoxynucleotide complementary to oxytocin mRNA blocks lactation in rats. J Clin Exp Endocrinol 1995a; 103:191-195.
68. Skutella T, Probst JC, Jirikowski GF. Antisense targeting of hypothalamic neuropeptides in vivo: biochemical, neuroendocrinological and behavioral aspects. Neuroendocrinology 1995b; 60:6.
69. Weis B, Zhou LW, Zhang SP, Quin ZH. Antisense oligodeoxynucleotides inhibits D2 dopamine receptor mediated behavior and D2 messenger RNA. Neuroscience 1993; 55:607-612.
70. Stanifer KM, Chien C-C, Wahlestedt C, Brown GP, Pasternak GW. Se-

lective loss of (opiod analgesia and binding by antisense oligodeoxynucleotides to a (opiod receptor. Neuron 1994; 12:805-810.
71. Baldwin HA, Britton KT, Koob GF. Behavioral effects of Corticotropin-releasing factor. In: Ganten D, Pfaff D, eds. Current Topics in Neuroendocrinology 1990; 10, Springer-Verlag: 1990:1-14.
72. Sutton RE, Koob GF, Le Moal M, Rivier J, Vale WW. Corticotropin-releasing factor (CRF) produces behavioral activation in rats. Nature 1981; 297:331-333.
73. Britton DR, Britton KT. A sensitive open field measure of anxiolytic drug activity. Pharmacol Biochem Behav 1981; 15:577-582.
74. Adamec RE, McKay D. The effects of CRF on a-helical CRF on anxiety in normal and hypophysectomised rats. J Psychopharmacology 1993; 7:346-354.
75. Berridge CW, Dunn AJ. CRF and restraint stress decrease exploratory behavior in hypophysectomised mice. Pharmacol Biochem Behav 1989; 34:517-519.
76. Britton KT, Lee G, Dana R, Risch SC and Koob GF, Activating and anxiogenic effects of corticotropin-releasing factor are not inhibited by blockade of the pituitary-adrenal system with dexamethasone. Life Science 1986; 39:1281-1286.
77. Davis M. The role of the amygdala in fear and anxiety. Ann Rev Neurosci 1992; 15:353-375.
78. Davis M, Rainnie D, Cassell M. Neurotransmission in the central amygdala related to fear and anxiety. Trends Neurosci 1994; 17:208-214.
79. Rassnik S, Heinrichs SC, Britton KT, Koob GF. Microinjection of a corticotropin-releasing factor antagonist into the central nucleus of the amygdala reverses anxiogenic-like effects in ethanol withdrawal. Brain Res 1992; 605:25-32.
80. Liebsch G, Landgraf R, Gerstberger R, Probst C, Wotjak CT, Engelmann M, Holsboer F, Montkowski A. Chronic infusion of a CRH1 receptor antisense oligodeoxynucleotide into the central nucleus of the amygdala reduced anxiety-related behavior in socially defeated rats. Regulat Peptides, 1995; 59:229-239.

CHAPTER 11

APPLICATION OF ANTISENSE STRATEGIES FOR THE ANALYSIS OF PROTEIN KINASE FUNCTIONS

Jesús Avila, Javier Díaz-Nido and Nieves Villanueva

INTRODUCTION

The reversible phosphorylation-dephosphorylation of proteins constitutes the main route by which extracellular signals produce their physiological responses in target cells and cellular functions are controlled. A network of protein kinases and phosphatases allows the transduction of signals from the extracellular medium to the nucleus where gene expression is modulated.[1] Thus, protein phosphorylation plays essential roles in the regulation of the cell cycle, cell survival and death as well as numerous cell type-specific functions.

Protein phosphorylation is usually studied by different in vitro assays. However, sometimes different protein kinases seem to modify in vitro the same sites on a particular protein substrate, with identical functional consequences. In these cases, it is important to analyze the actual contribution of each protein kinase to the phosphorylation of the analyzed protein in intact cells. Thus, the specific depletion of distinct protein kinases by antisense oligonucleotide treatments is of great value to analyze the phosphorylation of a protein substrate within cells.

The rationale for any antisense oligonucleotide treatment is to choose the appropriate oligonucleotide, deliver it into the cell, and select the time of treatment. Finally, the consequences of this treatment on the amount of the targeted protein kinase, on the phosphorylation of specific substrates for this kinase and on the possible cellular functions in which these substrates may be involved must be analyzed. Some rules

Antisense Strategies for the Study of Receptor Mechanisms, edited by Robert B. Raffa and Frank Porreca. © 1996 R.G. Landes Company.

must be followed to choose the appropriate antisense oligonucleotide. The oligonucleotide should have a size with enough length and the adequate sequence to form a stable duplex with the targeted mRNA. This can be calculated by measuring the D score (D score = ΔG duplex formation) as indicated by Stull et al.[2] On the other hand, the antisense oligonucleotide should be short enough to be delivered within the cell.[3]

Short antisense oligonucleotides (between 15 to 30 mer.) can be directly added to cultured cells. However, the delivery of larger oligonucleotides into cells demands the use of vectors which facilitate oligonucleotide transport. In particular, the utilization of liposomes constituted by cationic lipids seems of great efficiency.[4]

The entry of oligonucleotides and their stability inside the cells depend on the cell type. It has been indicated that the uptake depends on the size of the oligonucleotide but not on its sequence, since only the phosphate moiety appears to be essential for oligonucleotide uptake through specific cell surface receptors.[6,7] A higher uptake and a slower turnover of oligonucleotide[4] occur in neurons when compared to several non neural cells.[5] Additionally, modified oligonucleotides (particularly phosphorothioate oligodeoxynucleotides) have been used to prevent their degradation by nucleases.[5] Furthermore, the localization of the region complementary to the antisense oligonucleotide on the targeted mRNA should be closed to the initiation of the coding sequence for the protein.[5] The length of the treatment should allow the significant depletion of the targeted protein kinase. Thus, a time longer than the turnover of the targeted protein kinase is required. In any case, protein kinase depletion should be analyzed at different treatment times. Once the protein kinase is efficiently depleted, the analyses of the phosphorylation of specific substrates and other functional analyses should be performed. Finally, it is suitable to test if the synthesis of the depleted protein kinase takes place after removal of the added antisense oligonucleotide in order to rule out nonspecific toxic side effects of the treatment. In different works these previous criteria have been fulfilled to deplete different protein kinases in distinct cell types.

PROTEIN KINASE 2 CK2 (CASEIN KINASE 2)

Protein kinase CK2 is an oligomeric enzyme containing two catalytic subunits (α or α') and two regulatory subunits (β).[8] CK2 seems to perform a role during mitogenesis in cultured cells. Thus, antisense oligonucleotides to CK2 subunits inhibit cell growth stimulation by epidermal growth factor.[9] CK2 has also been implicated in neuritogenesis in cultured neuroblastoma cells. In these cells, a translocation of CK2 from the nuclei to the cytoplasm correlates with an increase in the phosphorylation of the microtubule-associated protein MAP1B and neurite extension.[10,11] The possible implication of CK2 in neuritogenesis has been confirmed upon CK2 depletion induced by an antisense oligonucleotide complementary to CK2 (α/α') mRNA.[12] The oligonucleotide was chosen according to the rules indicated in the Introduction, complementary to a DNA region close to the initiation of the coding sequence of the protein and with a suitable D score to form a stable oligonucleotide-mRNA hybrid. In antisense oligonucleotide-treated neuroblastoma cells, phosphorylation of MAP1B and neuritogenesis was prevented. Upon removal of antisense oligonucleotide, back phosphorylation of MAP1B by CK2 was observed and neurites were extended[12] (see Fig. 11.1). Additionally, a role for catalytic subunits of protein kinase CK2 for the nuclear localization of the kinase in neuroblastoma cells has been suggested by antisense oligonucleotide (complementary to CK2 (α/α') or β mRNA) treatment of these cells. Treatment with an antisense oligonucleotide complementary to CK2 β subunit mRNA did not alter the nuclear localization of CK2 α subunit. In contrast, treatment with an antisense oligonucleotide complementary to α subunit prevents the transport of CK2 β subunits to the nuclei of neuroblastoma cells[13] (see Fig. 11.2). Also antisense oligonucleotide

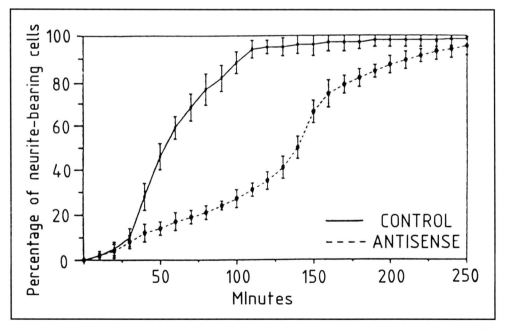

Fig. 11.1. Inhibition of neuritogenesis by casein kinase 2 antisense oligonucleotide is transient. The temporal course of neurite outgrowth in serum-free medium of N2A neuroblastoma cells either in the absence (control) or in the presence of the antisense oligonucleotide to CK2 (antisense) is shown. Reasumption of neurite growth after 2 h upon removal of the antisense oligonucleotide is accompanied by reappearance of immunoreactivity for antibodies to CK2. Reprinted with permission from: EMBO J 12:1633-1640. © Oxford Press.

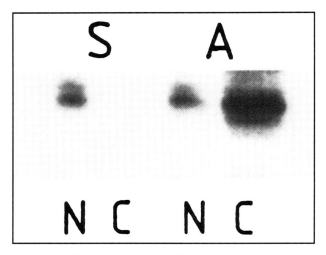

Fig. 11.2. Cell distribution of the CK2 β subunit in the absence of the CK2 α subunit. N2A neuroblastoma cells were treated with CK2 α sense (S), used as control, or antisense (A) oligonucleotides, and the presence of the β subunit in the nuclear (N) or cytoplasmic (C) fraction was determined by Western blotting using an antibody raised against CK2β subunit. Reprinted with permission from: Cell Mol Neurobiol; 14:407-414. © Plenum Publishing Corporation.

(complementary to CK α/α' subunits) treatment has been carried out in virus-infected cells. In cultured human (Hep-2) cells, oligonucleotide treatment has been done to characterize that CK2 is involved in the modification of a phosphoprotein of respiratory syncitial virus (RSV).[14] Oligonucleotide treatment of RSV-infected Hep2 cells results in a decrease of phosphate incorporation into the viral phosphoprotein (Fig. 11.3).

Therapeutical strategies using antisense oligonucleotides are now in their earlier developmental stages. One of these strategies has been postulated for one disorder in which CK2 appears to be involved, the cow disease termed theileriosis. This disease results from the infection of the parasite *T. parva* on bovine lymphocytes. In this process CK2 is dramatically increased in lymphocytes from infected blood, and it has been suggested that treatment with antisense oligonucleotides (complementary to the mRNA of CK2 catalytic subunits) could be used therapheutically.[15]

cAMP-DEPENDENT PROTEIN KINASE

cAMP-dependent protein kinase, or protein kinase A, is an oligomeric enzyme with the structure C_2R_2, being C and R the abbreviations for its catalytic and regulatory subunits, respectively. Antisense

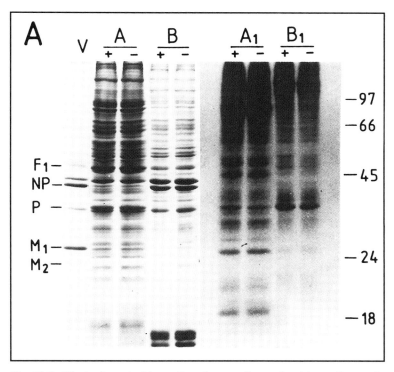

Fig. 11.3. Effect of casein kinase II antisense oligonucleotide on P protein phosphorylation and casein kinase II-like activity. HEp-2 cells were RS virus infected and 36 h postinfection were treated with sense (+) and antisense (-) oligonucleotides for 1 hour. Then, in the presence of the oligonucleotides were labeled with 25 µCi of ^{32}P- orthophosphate for 1 h. After the pulse, cytoplasmic extracts were prepared by sequential extraction with a buffer containing 0.5% NP-40 (A) or with a buffer containing 1% SDS (B). The protein concentrations were determined and 40 µg of protein from each extract were analyzed by gel electrophoresis. The proteins were visualized by staining with Coomassie blue panel A, (A) and (B), and those phosphorylated detected after autoradiography (A1) and (B1). Lane V corresponded to purified viral particles. Viral proteins are indicated in the left margin and the position of molecular weight markers in the right margin.

oligonucleotide (complementary to the mRNA of catalytic subunits) treatment was used to analyze the mechanism of desensitization of beta 2-adrenergic G protein-coupled receptors in different cell types (osteosarcoma, smooth muscle, epidermoid carcinoma and CHO cells).[16] The analysis was done in comparison to similar treatments performed with antisense oligonucleotides complementary to other kinases (protein kinase C and beta-adrenergic receptor kinase). The results obtained revealed distinctive cell-type-specific roles for cyclic AMP-dependent protein kinase, protein kinase C and beta-adrenergic receptor kinase in agonist-induced desensitization.[16]

However, more work has been done by using antisense oligonucleoide complementary to the regulatory subunits of protein kinase A. In this way, it has been shown that the RI alpha subunit, but not the RII beta subunit, of PKA controls serum dependency and entry into cell cycle of human mammary epithelial cells.[17] The expression of the RII beta subunit of protein kinase A could be essential for cAMP-induced growth inhibition in human leukemia cells.[18] Similar results have been reported in other cell types.[19] The previous suggested roles of the regulatory subunits of PKA in modulating cell proliferation or differentiation have been taken into account for possible therapeutic strategies in handling tumor cells. It has been described that reduction of RI alpha subunit expression results in growth inhibition of human epithelial cells transformed by TGFα, H-ras or erb B-2 genes.[20] Also it has been suggested that the addition of RI antisense oligonucleotide could be the basis for new approaches to prevent the proliferation of tumor cells.[21] These approaches have been suggested for an experimental gene therapy of human colon cancer.[22]

PKC

At least ten different isoforms of PKC representing the products of separate genes, have been described.[23] Those isoforms have been classified in classical (or conventional, novel (or non classical) and atypical isoforms, depending on their different requirements to show a full kinase activity.[23] PKC isoforms have been named according to the letters of Greek alphabet complemented in some cases with roman numerals. In this way conventional PKC are termed as α, βI, βII and γ isoforms, novel PKC include isoforms δ, ε, η and θ, and atypical PKC are ζ and λ isoforms. These different types of protein kinase C have been involved in different process of signal transduction mechanisms from the cell membrane to the nucleus resulting in changes in cell growth. Antisense treatments using oligonucleotides complementary to the different PKC types have been described. Examples are described below.

PKC α AND β ISOFORMS

Delivery of a 20-mer modified (phosphorothioate) oligonucleotide complementary to the sequence containing the initiation codon of the mRNA encoding protein kinase C-alpha inhibits the expression of this protein in cultured mouse epithelial cells. This also occurs in the liver of mice injected intraperitoneally with the antisense oligonucleotide, without affecting the expression of other PKC isozymes as delta, epsilon or zeta.[24]

In cultured cells such as hepatoma[25] or glioblastoma cells,[26] addition of the antisense oligonucleotide complementary to the mRNA of PKC alpha isoform results in an inhibition of cell growth. By antisense technology, a possible role of PKC alpha in the activation of phospholipase D has been suggested.[27] As a control, no effect on phospholipase D activation was observed when PKC beta was depleted in cultured cells.[27] However, a specific role for protein kinase C beta in the regulation of proliferation and differentiation of the human leukemia cell line HL-60 has been suggested by adding an antisense oligonucleotide complementary to the mRNA for PKC beta.[28]

OTHER PKC ISOFORMS

Incubation of mouse (NIH 3T3) cells with antisense oligonucleotides to PKC delta depletes this protein and specifically blocked platelet derived growth factor-stimulated mitogenesis in these cells; without affecting the mitogenesis promoted by other growth factors such as epidermal growth factor.[29] Regarding other functions for PKC isoforms in signal transduction mechanisms, a possible role of protein kinase zeta isoform in nerve growth factor induced differentiation of PC12 has been suggested by performing an experiment involving the addition of a 20-mer antisense oligonucleotide directed against the 5' coding sequence of PKC zeta mRNA. Upon addition of the oligonucleotide, a decrease in NGF-induced neurite outgrowth on PC12 cells was observed.[30]

PROLINE-DIRECTED PROTEIN KINASES (PDPK)

Several protein kinases phosphorylate serine or threonine residues followed by prolines. This modification could result in conformational changes affecting the functionality of the phosphorylated protein. Among these protein kinases, glycogen synthase kinase 3, cyclin-dependent kinases and mitogen-activated protein kinases are the best characterized.

GLYCOGEN SYNTHASE KINASE 3

Glycogen synthase kinase 3 (GSK3) is a family of two related proteins (GSK3α and GSK3β) that was first described as a modulator of glycogen metabolism, phosphorylating glycogen synthase and mediating insulin regulation of glycogen synthesis.[31,32] Recently, this kinase has also been implicated in the development of Alzheimer's disease since GSK3 is the same protein as tau-1 protein kinase, which modifies the microtubule-associated protein tau at some residues that are preferentially phosphorylated in the brain of Alzheimer's disease patients.[33] Interestingly, it has been observed that addition of an antisense oligonucleotide complementary to the mRNA of GSK3β attenuates the neurotoxicity induced by the beta amyloid peptide.[34]

CYCLIN-DEPENDENT PROTEIN KINASES (CDKS)

These PDPKs are holoenzymes composed of a regulatory subunit (cyclin) and a catalytic subunit, named cyclin-dependent kinase (or cdk).[35] The combination of different cyclins (identified with capital letters) with different cdk (termed with different numbers) results in different kinases that are active at different stages of the cell cycle.[35]

CDK1 AND CDK2

Injury to the endothelium of the rat carotid artery results in a progressive smooth muscle cell accumulation. During this process, a transient rise of cdk1 and cdk2 takes place. The muscle cell accumulation is inhibited by treatment with antisense oligonucleotides complementary to either mRNA of cdk-1 or cdk2.[36] Similar results were obtained by Morishita et al using oligonucleotides complementary to mRNA of cdk2 or cyclin B, suggesting a potential therapeutical value for these treatments.[37,38]

CYCLINS

The regulatory subunits of the cdk holoenzymes, cyclins, are selectively expressed at different times of the cell cycle.[35] Thus, depletion of a specific cyclin may result in the arrest of the cell cycle at a specific point. Using this approach, it has been shown that activation-induced T cell death is cell cycle dependent and is regulated by cyclin B.[39] In other experiments a role for cyclin D on DNA synthesis of human fibroblasts was shown.[40] In the same work it was indicated that treatment with antisense oligonucleotide to cyclin D accelerated DNA repair in these cells.[40] Furthermore, it has been shown that addition of cyclin D antisense-oligonucleotide results in the arrest of the cells in G1, in a process in which cdk4 and the product of the retinoblastoma also play a role.[41]

MITOGEN-ACTIVATED PROTEIN KINASES

A third family of PDPKs is constituted by the mitogen-activated protein kinases (MAP kinases) also referred to as extracellular signal regulated protein kinases (ERK).[42] Two members of this family, with relative molecular masses of 42k and 44k (ERK1 and ERK2) have been studied. Depletion of both MAP kinases blocked the ability of serum or insulin to stimulate DNA synthesis in fibroblasts or adipocytes. Also, the differentiation of fibroblasts into adipocytes was prevented in the presence of an antisense oligonucleotide complementary to both 42k and 44k MAP kinases.[43] Additionally, the use of an antisense oligonucleotide complementary to ERK2 mRNA has shown that such kinase is required for phorbol-stimulated proliferation of rat lymphoblasts.[44] The full activity of MAP kinases is obtained upon modification of the protein by a MAP kinase which is activated by phosphorylation with raf kinase.[45] Depletion of raf kinase by antisense oligo- nucleotide treatment inhibits bryostatin-1-mediated proliferation in hematopoietic cells,[46] abolishes insulin stimulation of DNA synthesis in hepatoma cells,[47] and prevents the mitogenic response of T-cells to interleukin-2.[48]

OTHER SERINE/THREONINE PROTEIN KINASES

The involvement of other serine/threonine protein kinases in different cellular functions has been analyzed by the antisense oligonucleotide technology. Several examples follow:

(1) A maize pollen-specific, calcium-dependent, calmodulin-independent protein kinase has been involved in the process of germination and pollen tube growth when an antisense oligonucleotide complementary for its mRNA was injected in maize.[49]
(2) A novel protein kinase, PTK1, has been described to be required for proliferation of human melanocytes, since cell growth is inhibited in the presence of a PTK1 antisense oligonucleotide.[50]
(3) A new serine/threonine protein kinase (DAP kinase) has been suggested to be a potential mediator of the gamma interferon-induced HeLa cell death.[51]
(4) Depletion of a protein kinase with homology to Drosophila *polo* gen product (a serine-threonine protein kinase) results in the arrest of DNA synthesis in NIH-3T3 cells.[52]

TYROSINE PROTEIN KINASES

The consequences of the depletion of different tyrosine protein kinases on cellular functions have also been studied using antisense oligonucleotides. Depletion of the intracellular tyrosine kinase megakaryocyte-associated tyrosine kinase (MATK) significantly inhibited megakaryocyte-precursor cell proliferation.[53] Treatment with antisense oligonucleotides complementary to bcr/abl protein tyrosine kinase decreases cell growth of leukemia cell lines.[54,55] Also, antisense treatment to deplete src tyrosine protein kinase expression inhibits proliferation of human leukemia cells.[56] Antisense oligonucleotide complementary to p59fyn tyrosine kinase inhibits the growth of bone marrow-derived macrophages.[57] This tyrosine kinase is also critical in TCR-mediated signalling.[58] Additionally, it has been found that p59fyn regulates the OKT3-induced calcium influx in Jurkat T cells. In another study using antisense oligonucleotides, it was indicated that the immunoglobulin-associated lyn tyrosine kinase is required for the antiimmunoglobulin mediated cell cycle arrest but it is not required for the signal leading to apoptosis.[60] A final example of the use of antisense oligonucleotides to study the role of a specific tyrosine protein kinase is that of the action of antisense oligonucleotides complementary to the stem cell tyrosine kinase 1 (STK-1) which results in the inhibition of the proliferation of progenitor-stem cells.[61]

PHOSPHATASES

Finally, it should be commented that the use of antisense oligonucleotides have also been used to study phosphatase functions. An example is the treatment with an antisense oligonucleotide complementary to the mRNA of the transmembrane protein tyrosine phosphatase named LAR, which is involved in the modulation of insulin receptor signaling in intact cells.[65]

CONCLUDING REMARKS

As indicated in the previous paragraphs, there are many reports indicating that the treatment with specific antisense oligonucleotides results in the inhibition of cell growth. Thus, some cautions should be taken to distinguish between sequence-dependent effects and nonspecific toxicity. In this way it should be pointed out that the introduction of two extra adenosine residues at the 3' end of an antisense oligonucleotide for the hematopoietically expressed homolog of cdc2 (cdk1) kinase could result in an increased cell toxicity.[62] Likewise, the addition of certain nucleotides to cultured cells could induce their death.[63] These effects may explain some results showing a similar effect of sense or antisense phosphorothrioate oligonucleotides on certain organ cultures.[64] Thus, the use of antisense oligonucleotide technique critically requires appropriate controls to distinguish specific from nonspecific effects. These should include the use of sense and mismatched oligonucleotide treatments as well as the observation of the reversion of an observed specific effect after removal of the antisense oligonucleotide.

REFERENCES

1. Karin M, Hunter T. Transcriptional control by phosphorylation: signal transmission from the cell surface to the nucleus. Curr Biol 1995; 5:747-755.
2. Stull RA, Taylor LA, Szoka FC. Predicting antisense oligonucleotide inhibitory efficacy: a computational approach using histograms and termodynamic indices. Nucleic Acid Res 1992; 20:3501-3508.
3. Caceres A, Kosik K. Inhibition of neurite polarity by tau antisense oligonucleotides in primary cerebelar neurons. Nature 1990; 343:461-463.
4. Akhtar S, Juliano RL. Cellular uptake and intracellular fate of antisense oligonucleotides. Trends Cell Biol 1992; 2:139-144.
5. Wahlestedt C. Antisense oligonucleotide strategies in neuropharmacology. Trends Pharmacol Sci 1994; 15:42-46.
6. Loke SL, Stein CA, Zhang X, Mori K, et al. Characterization of oligonucleotide transport into living cells. Proc Natl Acad Sci USA 1989; 86:3474-3478.
7. Akhtar S, Shoji Y, Juliano RL. In: Gene Regulation by antisense nucleic acids. Erikson RP and Izant J eds. Raven Press 1992: 133-145.
8. Issinger OG. Casein kinases: pleiotropic mediators of cellular regulation. Pharmacol Ther 1993; 59:1-30.
9. Pepperkok R, Lorenz P, Jakobi R, Ansorge W, Pyerin W. Cell growth stimulation by EGF: inhibition through antisense-oligodeoxynucleotides demonstrates important role of casein kinase II. Exp Cell Res 1991; 197:245-253.
10. Serrano L, Hernández MA, Díaz-Nido J, Avila J. Association of casein kinase II to microtubules. Exp Cell Res 1989; 181:263-273.
11. Díaz-Nido J, Serrano L, Méndez E, Avila J. A casein kinase II related activity is involved in phosphorylation of microtubule associated protein MAP1B during neuroblastoma cell differentiation. J Cell Biol 1988;

106:2057-2065.
12. Ulloa L, Diaz-Nido J, Avila J. Depletion of casein kinase II by antisense oligonucleotide prevents neuritogenesis in neuroblastoma cells. EMBO J 1993; 12:1633-1640.
13. Ulloa L, Díaz-Nido J, Avila J. Depletion of catalytic and regulatory subunits of protein kinase CK2 by antisense oligonucleotide treatment of neuroblastoma cells. Cell Mol Neurobiol 1994; 14:407-414.
14. Villanueva N, Navarro J, Méndez E, García-Albert I. Identification of a protein kinase involved in phosphorylation of the C-terminal region of human respiratory syncitial virus P protein. J Gen Virol 1994; 75:555-565.
15. Ole-Mo Yoi OK. Casein kinase II in theileriosis. Science 1995; 267:834-835.
16. Shih M, Malbon CC. Oligodexynucleotides antisense to mRNA encoding protein kinase A, protein kinase C and beta adrenergic receptor kinase reveal distinctive cell type specific roles in agonist induced desensitization. Proc Natl Acad Sci USA 1994; 91:12193-12197.
17. Tortora G, Pepe S, Bianco C et al. The RI alpha subunit of protein kinase A controls serum dependency and entry into cell cycle of human mammary ephitelial cells. Oncogene 1994; 9:3233-3240.
18. Tortora G, Budillon A, Yokozaki M et al. Retroviral vector mediated overexpresion of the RII beta subunit of the cAMP-dependent protein kinase induces differentiation in human leukemia cells and reverts the transformed phenotype of mouse fibroblasts. Cell Growth Diff 1994; 5:753-759.
19. Tortora G, Pepe S, Bianco C et al. Differential effects of protein kinase A subunits on Chinese hamster ovary cell cycle and proliferation. Int J Cancer 1994; 59:712-716.
20. Ciardiello F, Tortosa G, Pepe S et al. Reduction of RI alpha subunit of cAMP dependent protein kinase expression induces growth inhibition of human mammary epithelial cells transformed by TGF-alpha, c-Ha-ras and c-erb B-2 genes. Ann NY Acad Sci 1993; 698:102-107.
21. Cho CY, Clair T. The regulatory subunit of cAMP-dependent protein kinase as a target or chemotherapy of cancer and other cellular dysfunctional related diseases. Pharmacol Ther 1993; 60:265-288.
22. Bold RJ, Warren RE, Ishizuka J et al. Experimental gene therapy of human colon cancer. Surgery 1994; 116:189-195.
23. Tanaka C, Nishizuka Y. The protein kinase C family for neuronal signalling. Ann Rev Neurosci 1994; 17:551-567.
24. Dean NM, McKay R, Condon TP et al. Inhibition of protein kinase C alpha expresion in human A549 cells by antisense oligonucleotides inhibits induction of intracellular adhesion molecule 1 (I CAM-1) mRNA by phorbol esters. J Biol Chem 1994; 269:16416-16424.
25. Perletti GP, Smeraldi C, Porro D et al. Involvement of the alpha isoenzyme of protein kinase C in the growth inhibition induced by phorbol esters in MH1C1 hepatoma cells. Biochem Biophys Res Commun 1994; 205:1589-1594.
27. Balboa MA, Firestein BC, Godson C et al. Protein kinase C alpha medi-

ates phospholipase D activation by nucleotides and phorbol ester in Madin-Darby canine kidney cells stimulation of phospholipase D is independent of activation of polyphosphoinositide-specific phospholipase C and phospholipase A2. J Biol Chem 1994; 269:10511-10516.
28. Gamard CJ, Blobe GC, Hannun YA et al. Specific role for protein kinase C beta in cell differentiation. Cell Growth Differ 1994; 5:405-409.
29. Xu J, Rockow S, Kim S et al. Interferons block protein kinase C-dependent but not-independent activation of raf 1 and mitogen activated protein kinases and mitogenesis in NIH 3T3 cells. Mol Cell Biol 1994; 14:8018-8027.
30. Coleman ES, Wooten MW. Nerve growth factor-induced differentiation of PC12 cells employs the PMA insensitive protein kinase C-zeta isoform. J Mol Neurosc 1994; 5:39-57.
31. Woodget JR. Molecular clonning and expression of glycogen synthase kinase 3/factor A. EMBO J 1990; 9:2431-2438.
32. Woodget JR. A common denominator linking glycogen metabolism, nuclear oncogenes and development. Trends Biochem Sci 1991; 16:177-181.
33. Ishiguro K, Shiratsuchi A, Sato S et al. Glycogen synthase kinase 3β is identical to tau protein kinase I generating several epitopes of paired helical filaments. FEBS Letters 1993; 325:167-172.
34. Takashima A, Noguchi K, Sato K et al. Tau kinase I is essential for amyloid β protein induced neurotoxicity. Proc. Natl. Acad. Sci. USA 1993; 90:7789-7793.
35. Morgan DD. Principles of CDK regulation. Nature 1995; 374:131-134.
36. Abe J, Zhon W, Taguchi J et al. Suppression of neointimal smooth muscle cell accumulation in vivo by antisense cdc2 and cdk2 oligonucleotides in rat carotid artery. Biochem Biophys Res Commun 1994; 198:16-24.
37. Morishita R, Gibbons GH, Kaneda Y et al Pharmacokinetics of antisense oligodeoxyribonucleotides (cyclin B1 and CDC2 kinase) in the vessell wall in vivo: enhanced theraphetical utility for restenosis by HVJ liposome delivery. Gene 1994; 149:13-19.
38. Morishita R, Gibbons GH, Ellison KE et al. Intimal hyperplasia after vascular injury is inhibited by antisense cdk2 kinase oligonucleotides. J Clin Inv 1994; 93:1458-1464.
39. Fotedar R, Fla HJ, Crupta S et al. Activation-induced T-cell death is cell cycle dependent and regulated by cyclin B. Mol Cell Biol 1995; 15:932-942.
40. Pagano M, Theodoras AM, Ta SW et al. Cyclin D1-mediated inhibition of repair and replicative DNA synthesis in human fibroblasts. Genes Dev 1994; 8:1627-1639.
41. Tam SW, Theodoras AM, Shay JW et al. Differential expression and regulation of cyclin D1 protein in normal and tumor human cells: association with cdk4 is required for cyclin D1 function in G1 progression. Oncogene 1994; 9:2663-2674.
42. Boulton T, Nye SH, Robbins DJ et al. ERKS: a family of protein serine-threonine kinases that are activated and tyrosine phosphorylated in re-

sponse to insulin and NGF. Cell 1991; 65:663-675.
43. Sale EM, Atkinson PG, Sale GJ. Requirement of MAP kinase for differentiation of fibroblasts to adipocytes, for insulin activation of p90 56 kinase and for insulin or serum stimulation of DNA synthesis. EMBO J 1995; 14:674-684.
44. Lisboa C, Alemany S, Calvo V et al. Raf-1 and ERK2 kinases are required for phorbol 12, 13-dibutyrate-stimulated proliferation of rat lymphoblasts. ERK2 activation precedes Raf-1 hyperphosphorylation. Eur J Immunol 1994; 24:2764-2754.
45. Crews CM, Erikson RL. Extracellular signals and reversible protein phosphorylation: what to MEK of it all. Cell 1993; 74:215-217.
46. Carroll MP, May WS. Protein kinase C mediated serine phosphorylation directly activated raf-1 in murine hematopoietic cells. J Biol Chem 1994; 269:1249-1256.
47. Tornkvist A, Parpal S, Gustavson J et al. Inhibition of raf-1 kinase expression abolishes insulin stimulation of DNA synthesis in HA IIE hepatoma cells. J Biol Chem 1994; 269:13919-13921.
48. Riedel D, Brennscheidt U, Kiehntopf M et al. The mitogenic response of T cells to interleukin 2 requires raf-1 Eur J Immunol 1993; 23:3146-3150.
49. Estruch J, Kadwell S, Merlin E et al. Cloning and characterization of a maize polle-specific calcium-dependent calmodulin-independent protein kinase. Proc Natl Acad Sci USA 1994; 91:8837-8841.
50. Ezoe K, Lee ST, Strunc KM et al. PTK1, a novel protein kinase required for proliferation of human melanocytes. Oncogene 1994; 9:935-938.
51. Deiss LP, Feinstein E, Berissi H et al. Identification of a novel serine-threonine kinase and a novel 15-KD protein as potential mediators of the gamma interferon-induced cell death. Genes Dev 119; 9:15-30.
52. Hamanaka R, Maloid S, Smith MR et al. Cloning and characterization of human and murine homologues of Drosophila polo serine-threonine kinase. Cell Growth Diff. 1994; 5:249-257.
53. Aurahama S, Jiang S, Ota S et al. Structural and functional studies of the intracellular tyrosine kinase MATK gene and its translated product. J Biol Chem 1995; 270:1833-1842.
54. Okabe M, Kunieda Y, Miyagishima T et al. BCR/ABL oncoprotein-targeted antitumor activity of antisense oligonucleotides complementary to bcr/abl mRNA and herbimycin A, an antagonist of protein tyrosine kinase: inhibitors effects on in vitro growth of Ph1-positive leukemia cells and BCR/ABL oncoprotein-associated transformed cells. Leuk Lymphoma 1993; 10:307-316.
55. Skorski T, Kanakaraj P, Ku D et al. Negative regulation for p120GAP GTPase promoting activity of p21 bcr/abl: implication for RAS-dependent Philadelphia chromosome positive cell growth. J Exp Med 1994; 179:1955-1965.
56. Kitanaka A, Waki M, Kamano H et al. Antisense src expression inhibits proliferation and erythropoietin-induced erythroid differentiation of K562 human leukemia cells. Biochem Biophys Res Commun 1994; 201:1534-1540.

57. Li Y, Chen B. Differential regulation of fyn-associated protein tyrosine kinase activity by macrophage colony-stimulating factor (M-CSF) and granulocyte macrophage colony-stimulating factor (GM-CSF). J Leukoc Biol 1995; 57:484-490.
58. Lee S, Shaw A, Maher SE et al. P59FYN tyrosine kinase regulates p561ck tyrosine activity and early TCR-mediated signalling. Int Immunol 1994; 6:1621-1627.
59. Rigley K, Slocombe P, Prondfoot K et al. Human p59fyn regulates OKT3-induced calcium influx by a mechanism distinct from PIP2 hydrolysis in Jurkat T cell. J Immunol 1995; 154:1136-1145.
60. Schevermann RH, Racila E, Tucker T et al. Lyn tyrosine kinase signals cell cycle arrest but not apoptosis in B-lineage lymphoma cells. Proc Natl Acad Sci USA 1994; 91:4048-4052.
61. Small D, Levenstain M, Kim E et al. STK-1 the human homology of Flk/FlT3 is selectively expressed in CD34+ human marrow cells and is involved in the proliferation of early progenitor/stem cells. Proc Natl Acad Sci USA 1994; 91:459-463.
62. Ehrlich G, Patinkin D, Ginzberg D et al. Use of partially phosphothioated antisense oligodeoxynucleotides for sequence-dependent modulation of hematipoiesis in culture. Antisense Rev Dev 1994; 4:173-183.
63. Wakade AR, Przywara DA, Palmer KC et al. Deoxynucleoside induces neuronal apoptosis independent of neurotrophic factors. J Biol Chem 1995; 270:17986-17992.
64. Durbeej M, Soderstrom S, Ebendal T et al. Differential expression of neurotrophin receptors during renal development. Development 1993; 119:977-989.
65. Kulas DT, Zhang WR, Goldstein B et al. Insulin receptor signalling is augmented by antisense inhibition of the protein tyrosine phosphatase LAR. J Biol Chem 1995; 270:2435-2438.

= CHAPTER 12 =

APPLICATION OF ANTISENSE TECHNOLOGY FOR STUDYING THE FUNCTIONAL ROLE OF TAU PROTEINS IN NEURAL PLASTICITY

Maurizio Memo

INTRODUCTION

The concept of neural plasticity is often referred to a specific cellular compartment, i.e., the synapse, in which several factors involved in cell-to-cell communication may continuously modify their kinetic and functional properties to integrate external stimuli into cell function. Plasticity can also be viewed in a more general fashion as the ability of the cells to move and/or change morphology. These effects range from playing a crucial role in ensuring the selection and survival of developing neurons to relatively delicate effects in the number, length and strength of synaptic contacts after environmental or pharmacological stimuli. In any case, plasticity is a regulated phenomenon that involves extracellular factors, including neurotransmitters and neurotrophins, and intracellular targets, including cytoskeletal proteins.

Differentiation of the nervous system provides an example that argues persuasively for a sophisticated neural plasticity.[1] At the very early stage of embryogenesis, neuroblasts arise at particular locations at specific times. Such cells generally lack axonal or dendritic projections. Neuroblasts migrate extensively in the central nervous system to reach specific destinations where they stop. At its destination the migrating cell sprouts and elaborates one or more axonal and dendritic projections. At the leading edge of the elongating axon is the highly motile structure known as growth cone. As the growth cone moves outward, the cell body stays put and the axon elongates due in part to polymerization of tubulin into microtubules that give the axon its rigidity. In cell culture as well as in vivo, growth cone-mediated axon elongation is stimulated by many substances. These substances would coordinately change the pattern of expression of a number of genes to make several proteins available for building new intracellular structures. Among the most relevant proteins involved in modifying cell morphology are those required to make the cytoskeleton of the cell, i.e., the microtubule.

Microtubules are composed of subunits that are heterodimers of α and β tubulin monomers and are characterized by an intrinsic dynamic instability. There are a number of parameters that determine the stability of microtubules. Among these is the concentration of free tubulin: a high concentration favors continued growth, and a low concentration allows GDP cup to form at the end, causing the microtubules to depolymerize.[2] Another parameter is the rate of hydrolysis of GTP to GDP: a slow rate favors continued microtubule growth and a fast rate favors shrinkage.[3] Microtubule stability is also determined by the availability of the microtubule-associated protein (MAP) tau. Tau is a cytoskeleton-associated, neuronal protein promoting microtubule assembly and neurite polarization.[4] It is thought to make short crossbridges between axonal microtubules to prevent depolymerization. In fact, tau proteins are a class of low molecular mass proteins expressed in neural cells which are closely related by amino acid composition. Since tau proteins are thought to be encoded by a single gene, it is considered likely that tau heterogeneity arises via both differential mRNA processing and posttranslational modifications. The most striking feature of the primary structure of tau as predicted from molecular cloning is a stretch of 31 or 32 aminoacids that is repeated three or four times in the carboxy-terminal half of the molecule. Additional isoforms exist which contain 29 or 58 amino acid insertions in the amino-terminal region in conjunction with three or four repeats, giving rise to a total of six different tau isoforms.[5] The longest tau isoform, which contains four-repeats and the 58 amino acid insertion, is found in adult but not in fetal human brain, whereas the shortest tau isoform, which contains three repeats, is found in fetal brain and in a restricted populations of neurons in adult brain. In particular, neurons expressing mRNAs encoding three-repeat containing tau isoforms are the pyramidal cells throughout all layers of the cerebral cortex and both the hippocampal granule and pyramidal cells. In contrast with granule cells, pyramidal cells in hippocampus and in cortex also express the mature tau isoform.[6,7]

Thus, in the adult human brain, there are neurons that express selectively the mature tau isoform, neurons that express selectively the fetal tau isoform, and neurons that express both.

TAU PROTEINS IN NEURONS DEVELOPING IN VITRO

As an experimental paradigm for studying the involvement of tau proteins in neural plasticity we used primary cultures of cerebellar granule cells from neonatal rats. When cultured in standard conditions, these cells show short life-span, do not branch and die after 8-10 days. Survival and differentiation can be accomplished by exposing these cells to the excitatory neurotransmitter glutamate. Alternatively, neural differentiation can be promoted by culturing the cells in media containing high potassium concentrations. Because of the homogeneity of this cell preparation, it is believed that both glutamate treatment and high potassium activate a genetic program leading to morphologically and phenotipically established neurons, in a similar manner.[8,9] We found that expression of tau proteins in primary cultures of cerebellar granule cells is a developmentally regulated process affecting different steps of tau processing, including RNA splicing and translation.[10]

Changes in tau RNA splicing are clearly demonstrated by PCR data showing the switching on of the exon containing the four internal repeat at 6 days in vitro (DIV) and the switching off of the mRNA containing three internal repeats after DIV 12. In the rat, the fetal tau isoform is expressed in early postnatal brain while the mature isoform is expressed in mature brain.[11] The shift to mature tau expression is approximately at postnatal day 8, which is the age of the rat used in our study. Since at that time cerebellar granule cells are still undifferentiated, it was not surprising to find that in the first period of culturing (DIV 1 to 4) these cells express preferentially, if not exclusively, the fetal isoform of tau. As shown by immunocytochemistry and Western blot analysis with anti-tau antibodies,[10] at this time period cells express predominantly a panel of low MW proteins that are localized in both soma and growing neurites. This is in line with previous observations in PC12 cells after NGF-induced differentiation.[12] In our experimental paradigm, cerebellar granule cells were allowed to differentiate for several days in culture, a period of time sufficient to permit the development of neurites and possibly to allow synaptic contacts. This in vitro differentiation was responsible for the activation of splicing mechanism resulting at 6 DIV in switching on of the four repeats mature tau isoform. It is tempting to speculate that after 4-6 days of culture, a significant amount of neurons established synapses allowing factor(s) from the target cells to address a series of retrograde, coordinated signals both to stop neurite growth and to differentiate the impinging axons. Part of this crucial stage of cell maturation might be the switch in the pattern of tau expression.

TAU ANTISENSE IN NEURONS DEVELOPING IN VITRO

Antisense strategy is based on the assumption that a given antisense oligonucleotide forms RNA-DNA hybrids with the endogenous sense sequence, thus reducing efficiency of the translation, stability or transport of the mRNA concerned.[13,14] For our purpose, an oligonucleotide complementary to a 26 nucleotide sequence comprising the ATG translation initiation codon of tau gene, and its exact inverse complement, the sense oligonucleotide, were synthesized and added to the cultured neurons at 2 DIV. Treatment was repeated for three consecutive days. The experimental protocol used in the present study, including oligonucleotide sequence, concentration, and time of exposure, was similar to that established by Caceres and Kosik for demonstrating the inhibition of neurite polarity by tau antisense oligonucleotides in primary cultures of cerebellar macroneurons.[15] We found that addition of different concentrations (from 10 μM to 50 μM) of tau oligonucleotide antisense to cultures of cerebellar granule cells at 2 DIV for three consecutive days induced a sustained reduction both of TAU-2 immunoreactivity and neurite growth. The maximal effect was obtained with 50 μM tau oligonucleotide antisense and resulted in a complete disappereance of TAU-2 immunoreactivity associated with no changes in MAP-2 immunoreactivity. A morphological study showed that tau antisense treatment decreases the number of cells with, and the length of neurites, without affecting cell survival. Interestingly, the same time of exposure and range of concentrations did not alter both TAU-2 immunoreactivity and neurite extension in cultures of cerebellar granule cells at 12 DIV, a time at which they are morphologically mature with established branches and neurites (Table 12.1). The greater sensitivity to tau antisense oligonucleotides of differentiating cells has been previously shown by Caceres and Kosik in cerebellar neurons[15] and by Hanemaaijer and Ginzburg in PC12 treated with nerve growth factor.[12]

Although the effect of the antisense was readily detectable, an attempt was made to demonstrate that the oligonucleotide entered the cells. The ^{32}P-labeled tau antisense oligonucleotide was found within the neurons 15 min after application. The amount of radioactivity incorporated by the cells increased with time reaching a plateau at 1 h. An acrylamide gel of the DNA/RNA cell extracts at the same time intervals revealed that nearly all the labeled oligonucleotide inside the cells remained intact after passage through plasma membranes. The pattern of incorporation of the labeled sense oligonucleotide by the cells was similar to that of the antisense and both were not affected by glutamate treatment.

TAU IN MATURE AND DEGENERATING NEURONS

The following set of experiments was performed in cerebellar granule cells at 12 DIV. At this stage of development, cerebellar granule cells are morphologically differentiated and show, contrary to earlier stages

Table 12.1. Effect of different concentrations of tau oligonucleotide antisense on tau immunoreactivity

Concentration	Tau immunoreactivity	
	DIV 2	DIV 12
–	37 ± 8	39 ± 7
5 µM	36 ± 5	39 ± 4
25 µM	17 ± 7*	37 ± 3
50 µM	5 ± 2*	15 ± 2*

Cells were pretreated with vehicle or tau antisense oligonucleotide for 24 hr. Semiquantitative analysis of TAU-2 immunoreactivity was done by Image Analyis program Genias on Magiscan System by Joyce-Loebl Ltd. Values are the means ± SEM of specific densities expressed as ID/area x 10^3 of at least 100 cells taken from 3 different dishes in two separate experiments.
* $p < 0.01$ vs the corresponding untreated controls. DIV, days of culturing in vitro. Experimental details are in ref. 18.

of maturation, a peculiar vulnerability to the neurotoxic effects induced by glutamate.[16] As reported above, these cells express different species of mRNAs deriving by alternative splicing of the mRNA encoding tau proteins. We found that short-term exposure of cerebellar granule cells to increasing concentrations of glutamate resulted in a dose-dependent increase of tau immunoreactivity. This effect was detectable 2 h after the glutamate pulse with two distinct anti-tau antibodies, TAU-2 and Alz-50.[17] Since both TAU-2 and Alz50 antibodies do not discriminate between phosphorylated and nonphosphorylated form of tau, it was unclear whether or not glutamate activates specific protein kinases in cerebellar granule cells to increase the phosphorylation state of tau proteins.

We were then interested in evaluating whether the increase in tau immunoreactivity detected in cerebellar granule cells after the glutamate pulse was a consequence of an increased tau gene transcription. We found that glutamate increases in a time- and concentration-dependent fashion both mature and fetal tau mRNA species.[18] The effects of glutamate on tau immunoreactivity and mRNA levels might be causally and sequentially related. The correlation is suggested on the bases of similar dose-response and time-course curves.

Doses of glutamate that do not induce neuronal death were able to increase both mRNA levels of and immunoreactivity to tau protein. These findings suggest that the increased tau expression detectable after high doses of glutamate is not a general response to degenerative stimuli and raises the question whether or not tau accumulation may reflect an early stage of degeneration. Alternatively, tau expression may represent a sign of a potential neuron vulnerability. In line with this is the work by Al-Ghoul and Miller, who showed that tau immunoreactivity is a marker for neurons susceptible to naturally occurring cell death during brain development,[19] and that by Mattson, who demonstrated that in a heterogeneous cell preparation, i.e., primary culture of hippocampal neurons, glutamate induced degeneration only in those neurons showing changes in tau immunoreactivity.[20] Finally, high levels of transcripts encoding different isoforms of tau protein are found specifically in the bodies of the most vulnerable cell types of the brain, the pyramidal cells of cerebral cortex and hippocampus.[7]

In this regard it should be mentioned that tau proteins are also involved in the formation of one of the neuropathological features of Alzheimer's disease: the neurofibrillary tangles (NFTs). There is increasing evidence supporting the hypothesis that one of the major steps in NFT formation is impairment of the intracellular mechanisms regulating tau expression. In comparison to those found in the majority of neurons from adult brain, tau proteins in NFTs are abnormally located —soma vs axons—, aberrantly phosphorylated, and translated by transcripts that are time-inappropriate-fetal vs mature tau mRNA isoforms.[21-23] As a possible result of these combinatory processes is the formation of insoluble so-called paired helical filaments (PHF), that in turn might be responsible for aggregation of other proteins to form NFTs. The above mentioned alterations of tau gene processing likely involve both transcriptional and posttranscriptional changes revealing a complex modification of the cell machinery devoted to cytoskeleton formation and stabilization.

TAU ANTISENSE IN DEGENERATING NEURONS

There is an emerging consensus that glutamate, through the activation of specific glutamate receptor subtypes, activates a series of immediate early genes triggering a long-lasting transcriptional program which may result in the regulation of the expression of various proteins.[24,25] Particularly, it has been previously established that stimulation of NMDA-selective glutamate receptors that are present in primary culture of cerebellar granule cells results in the induction of a number of immediate early genes, including c-*fos*, c-*jun*, *jun*-B and *zif*/268.[24] The protein products of these genes have been postulated to function as nuclear third messengers in coupling receptor stimulation to long term phenotypic changes in neurons. It is tempting to speculate that tau gene is one of the target genes that are regulated by these transcriptional factors. Nevertheless, the functional contribution of individual proteins in processing the glutamate signal to induce neuronal death is still unknown. We thus investigated the possible involvement of tau proteins in the neurotoxic process activated by glutamate using the oligonucleotide antisense strategy. We found that preincubation of cerebellar granule cells with a specific tau antisense oligonucleotide resulted in an inhibition of the glutamate-induced both tau mRNA levels (Fig. 12.1) and tau immunoreactivity (Fig. 12.2).[18] Specificity of this effect was proved since pretreatment of the cells with the sense oligonucleotide did not change the ability of glutamate to increase tau immunoreactivity. In addition, we found that the glutamate-induced increase of tau immunoreactivity detectable by Western blot analysis was prevented by exposing the cells to a tau antisense oligonucleotide. Functionally, the inhibition of tau neosynthesis by tau antisense oligonucleotide treatment resulted in a significant decrease of the sensitivity of the neurons to neurotoxic concentrations of glutamate (Fig. 12.3).[18] This protective effect was not due to an impairment of the ability of glutamate to increase the intracellular calcium concentration, suggesting that the sustained and abnormally high glutamate-induced increase of $[Ca^{2+}]i$ is not always related with neuronal death.

CONCLUSION

Central nervous system development, function, plasticity and perhaps degeneration depend on the coordinated transcriptional modifications of sets of genes encoding proteins that are relevant to particular neuronal function. In vitro studies with morphologically homogeneous

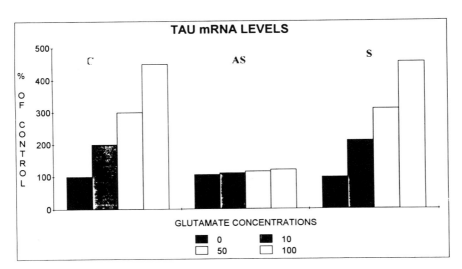

Fig. 12.1. Effect of tau oligonucleotide antisense (AS) on glutamate-induced tau mRNA levels. C, control; S, oligonucleotide sense. Bars represent experimental data obtained with increasing concentrations of glutamate (10 µM, 50 µM, 100 µM). Experiments were done in primary cultures of cerebellar granule cells from neonatal rats. Cells were exposed to 50 µM glutamate for 15 min. Tau oligonucleotide antisense (or sense where indicated) was added to the cultures 30 min before the glutamate pulse. Tau mRNA levels were measured by a quantitative RT-PCR two hours after the glutamate pulse. Experimental details are in ref. 18.

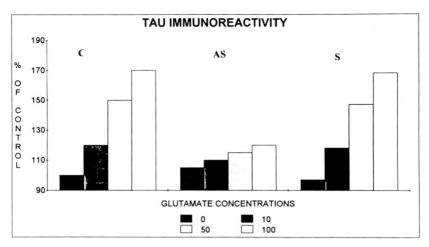

Fig. 12.2. Effect of tau oligonucleotide antisense (AS) on glutamate-induced tau expression. C, control; S, oligonucleotide sense. Bars represent experimental data obtained with increasing concentrations of glutamate (10 μM, 50 μM, 100 μM). Experiments were done in primary cultures of cerebellar granule cells from neonatal rats. Cells were exposed to 50 μM glutamate for 15 min. Tau oligonucleotide antisense (or sense where indicated) was added to the cultures 30 min before the glutamate pulse. Tau protein levels were detected by immunocytochemistry, using Alz-Antisense Technology and the Functional Role of Tau Proteins in Neural Plasticity50 tau antibody, two hours after the glutamate pulse. Experimental details are in ref. 18.

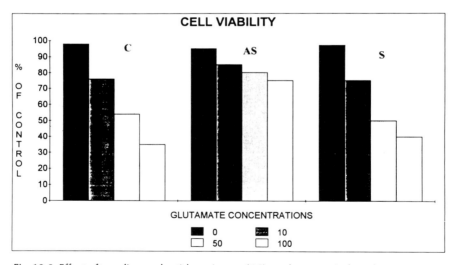

Fig. 12.3. Effect of tau oligonucleotide antisense (AS) on glutamate-induced neurotoxicity. C, control; S, sense oligonucleotide. Bars represent experimental data obtained with increasing concentrations of glutamate (10 μM, 50 μM, 100 μM). Experiments were done in primary cultures of cerebellar granule cells from neonatal rats. Cells were exposed to 50 μM glutamate for 15 min. Tau oligonucleotide antisense (or sense where indicated) was added to the cultures 30 min before the glutamate pulse. Viability was measured 24 hours after the glutamate pulse. Experimental details are in ref. 18.

populations of neuronal cells in primary culture provide experimental avenues for the elucidation of the regulatory mechanisms for genetic programs that are operative in various aspects of neuronal life. Because cytoskeletal proteins are the arbiters of cellular form and structure, it would be expected that any reactive or normally occurring morphological changes in neurons would be mediated through effects on cytoarchitecturally important proteins. The present data indicate that neosynthesis of the cytoskeleton-associated tau protein is a crucial step in the cascade of events promoted by glutamate leading to both neurite growth and neurodegeneration. Thus, regulation of tau synthesis might represent a common pattern by which glutamate may induce axonal maturation in developing neurons and neurodegeneration in selected, vulnerable, differentiated neurons.

This view is supported by morphological observations in Alzheimer's brain of numerous contorted processes from cell bodies of NFT-bearing neurons and supernumerary basilar dendrites on hippocampal pyramidal cells.[26] All these phenomena may be part of an uncontrolled growth response of established neurons. Since the inhibition of tau synthesis does not completely prevent but only decreases the neuronal sensitivity to the glutamate-induced cell death program, it is allowed to speculate that accumulation of tau in response to glutamate represents a molecular risk factor for neurodegeneration which contributes, together with others, to lower the safety margin of neurons to excitotoxin-induced injury.

REFERENCES

1. Aguayo A, Raff M (eds). Development. Curr Op Neurobiol 1995; 5:1-133.
2. Okabe S, Hirokawa N. Microtubule dynamics in nerve cells: analysis using microinjection of biotinylated tubulin into PC12 cells. J Cell Biol 1988; 107:651-664.
3. Carlier MF. Role of nucleotide hydrolisis in the polymerization of of actin and tubulin. Cell Byophis 1988; 12:105-117.
4. Hirokawa N, Shiomura Y, Okabe S. Tau proteins: the molecular structure and the mode of binding on microtubules. J Cell Biol 1989; 107:1449-1459.
5. Goedert M. Tau protein and the neurofibrillary pathology of Alzheimer's disease. Trends Neurosci 1993; 16:460-465.
6. Goedert M, Spillantini MG. Molecular neuropathology of Alzheimer's disease: in situ hybridization studies. Cell Mol Neurobiol 1990; 10, 159-174.
7. Goedert M, Sisodia SS, Price DL. Neurofibrillary tangles and β-amyloid deposits in Alzheimer's disease. Curr Op Neurobiol 1991; 1:441-447.
8. Balazs R, Hack N, Jorgensen OS, Cotman CW. N-Methyl-D-Aspartate promotes the survival of cerebellar granule cells: pharmacological characterization. Neurosci Lett 1989; 101: 241-246.

9. Gallo V, Kingsbury A, Balazs R, Jorgensen OS. The role of depolarization in the survival of cerebellar granule cells in culture. J Neurosci 1987; 7:2203-2213.
10. Pizzi M, Valerio A, Belloni M, Arrighi V, Alberici A, Liberini P, Spano PF, Memo M. Differential expression of fetal and mature tau isoforms in primary cultures of cerebellar granule cells during differentiation. Mol Brain Res 1995; 4:38-44.
11. Kosik KS, Orecchio LD, Bakalis S, Neve R. Developmentally regulated expression of specific tau sequences. Neuron 1989; 2:1389-1397.
12. Hanemaaijer R, Ginzburg I. Involvement of mature tau isoforms in the stabilization of neurites in PC12 cells. J Neurosci Res 1991; 30:163-171.
13. Eguchi Y, Itoh T, Tomizawa, J-I. Antisense RNA. Annu Rev Biochem 1991; 60: 631-652.
14. Heidenreich O, Kang S-H, Xu X, Nerenberg M. Application of antisense technology to therapeutics. Mol Medicine Today 1995:128-133.
15. Caceres A, Kosik KS. Inhibition of neurite polarity by tau antisense oligonucleotides in primary cerebellar neurons. Nature 1990; 343: 461-463.
16. Resnik A, Hack N, Boer GJ, Balazs R. Growth conditions differentially modulate the vulnerability of developing cerebellar granule cells to excitatory amino acids. Brain Res 1994; 655:222-232.
17. Pizzi M, Valerio A, Ribola M, Spano PF, Memo M. A tau antisense oligonucleotide decreases neuron sensitivity to excitotoxic injury. NeuroReport 1993; 4: 823-826.
18. Pizzi M, Valerio A, Arrighi V, Galli P, Belloni M, Ribola M, Alberici A, Spano PF, Memo M. Inhibition of glutamate-induced neurotoxicity by a tau antisense oligonucleotide in primary culture of rat cerebellar granula cells. Eur J Neurosci 1995; 7:1603-1616.
19. Al-Ghoul WM, Miller MW. Transient expression of Alz-50 immunoreactivity in developing rat neocortex: a marker for naturally occurring neuronal death? Brain Res 1989; 481: 361-367.
20. Mattson MP. Antigenic changes similar to those seen in neurofibrillary tangles are elicited by glutamate and Ca^{++} influx in cultured hippocampal neurons. Neuron 1990; 2:105-117.
21. Brion JP, Passareiro H, Nunez J, Flament-Durand J. Immunological determinants of tau proteins are present in neurofibrillary tangles of Alzheimer's disease. Arch Biol 1985; 95:229-235.
22 Goedert M, Wischik C, Crowther R, Walker J, Klug A. Cloning and sequencing of the cDNA encoding a core protein of the paired helical filament of Alzheimer's disease: identification as the microtubule-associated protein tau. Proc Natl Acad Sci USA 1988; 85:4051-4055.
23. Mandelkow E-M, Mandelkow E. Tau as a marker for Alzheimer's Disease. Trends Biol Sci 1993; 18:480-483.
24. Szekely AM, Costa E, Grayson DR. Transcriptional program coordination by NMDA-sensitive glutamate receptor stimulation in primary culture of cerebellar neurons. Mol Pharmacol 1990; 38:624-633.

25. Memo M, Bovolin P, Costa E, Grayson DR. Regulation of g aminobutyric acid-A receptor subunit expression by activation of N-methyl-D-aspartate-selective glutamate receptor. Mol Pharmacol 1991; 39:599-603.
26. Kosik KS. Tau proteins and Alzheimer's disease. Curr Opin Cell Biol 1990; 2:101-104.

CHAPTER 13

Behavioral Assessment of Antisense Oligonucleotides Targeted to Messenger RNAs of Genes Associated with Alzheimer's Disease

David P. Binsack, Sudhir Agrawal and Charles A. Marotta

INTRODUCTION

The National Institute on Aging has indicated that Alzheimer's Disease (AD) affects 10 percent of the population over the age of 65, and 50 percent of the population age 85 or older. Thus, this disorder is a major public health concern. The progressive clinical stages of AD not only incapacitate afflicted individuals and hasten death, but also place a substantial personal and economic burden on caretakers and society. While there are numerous therapeutic initiatives aimed at AD at the present time, most are directed towards downstream events, such as replacement of susceptible neurotransmitter systems in brain or administration of neurotrophic factors to revitalized degenerating neurons. By contrast, the application of antisense oligonucleotides (ASOs) targeted to the messenger RNAs (mRNAs) of genes that are integral to the molecular pathogenesis of AD provides a direct and specific

---

Antisense Strategies for the Study of Receptor Mechanisms,
edited by Robert B. Raffa and Frank Porreca. © 1996 R.G. Landes Company.

approach to interrupting certain of the neuropathologic. ASOs have structural features that may allow this class of compounds to be adapted to a wide variety of genes associated with neurodegenerative diseases irrespective of their linear sequence.

AD victims suffer declining memory and cognition, most likely related to the progressive loss of cholinergic neurons and then, subsequently, other neurotransmitter systems may be compromised. Strategies for therapeutic interventions need to take into account potential side effects that may exacerbate the specific symptomatology of the disorder. During the course of our program to design and develop antisense compounds for AD, we recognized the need to establish a systematic means for behavioral assessment of previously uncharacterized phosphorothioate oligodeoxynucleotides. This chapter describes behavioral assays used to assess components of learning and memory in rats exposed to oligodeoxynucleotide phosphorothioates. The paradigms were chosen to reflect, as much as is possible in the limitations of current animal models, certain of the symptoms experience by victims of AD.

ALZHEIMER'S DISEASE

In AD there is a decline in memory and other cognitive functions in comparison with the patient's previous level of function as determined by a history of decline in performance and by abnormalities noted from clinical examination and neuropsychological tests.[1] In definite dementia there is an acquired persistent impairment of intellectual functions with compromise in at least three of the following spheres of mental activity: language, memory, visuospatial skills, emotion or personality and cognition (abstraction, calculation, judgment, and executive function).[2] According to NINCDS-ADRDA Work Group criteria the definite diagnosis of AD depends on the combination of the clinical presentation of progressive dementia and histopathologic documentation of neuropathologic deterioration. While there is no uniform agreement on the quantitative and qualitative neurodegenerative changes in brain that define the diagnosis of AD, most diagnostic criteria include amyloid-containing senile plaques and neurofibrillary tangles (NFTs),[3,4] however, considerable weight has been given to the senile plaques, with exclusion of NFTs.[5] It has been observed that certain cases that come to postmortem examination have numerous plaques in the cortex without significant numbers of cortical tangles, which may be absent or sparse.[6]

AMYLOID PRECURSOR PROTEIN AND β-AMYLOID

Neuroanatomically, microscopic examination of the AD brain at autopsy reveals large numbers of neuritic plaques consisting of a central core of amyloid surrounded by a cluster of dystrophic degenerating neurites and activated glial cells.[7] Although the plaques may be distributed throughout several areas in the brain, the heaviest concentrations are found in the entorhinal cortex, the dentate gyrus and CA1 subfield of the hippocampus, and in the lateral nucleus of the amygdala; lesser concentration are seen in the dorsolateral frontal cortex, orbital frontal cortex, posterior parietal cortex and posterior temporal cortex. Although plaques have been shown to contain a number of proteins in minor amounts (e.g., α_1-antichymotrypsin, apolipoprotein E, lysosomal proteases, and apolipoprotein E (*vide infra*), by far the most prominent component is β-amyloid. The latter is a 4.2 kDa peptide containing 39-43 amino acids that can adopt β-pleated sheet conformation and which is derived from the amyloid precursor protein (APP). The APP occurs in three major isoforms (APP_{770}, APP_{751}, APP_{695})[8] that arise from alternative RNA splicing of a primary transcript located on the long arm of chromosome 21.[9]

The importance of β-amyloid to the neuropathogenesis, and in some cases to the etiology, of AD is supported by diverse experimental approaches. β-amyloid has been shown to have neurotoxic potential in tissue culture studies.[10] Acute addition of β-amyloid to cultures causes a trophic response; however, upon prolonged incubation β-amyloid forms insoluble aggregates with neurotoxic properties.[11,12] In vivo administration of β-amyloid has been reported to induce neurodegenerative changes in animals.[13,14]

Data derived from genetic studies provide further support for the importance of β-amyloid to AD pathology. In certain families with heritable forms of AD the disease is linked with mutations adjacent to the β-amyloid region of the APP.[15-19] A double mutation is found at APP_{770}670/671 upstream of the β-amyloid.[15,16] Single mutations of APP_{770}717 just beyond the β-amyloid C-terminus are associated with AD in several families.[17,18] Presumably the mutations interfere with the normal processing of the APP so as to increase the production of β-amyloid. Consistent with this hypothesis was the demonstration that genetically modified cultured cells that express the APP_{770}670/671 mutation secrete elevated levels of β-amyloid in vitro.[19,20] In hereditary cerebral hemorrhage with amyloidosis of the Dutch type, there occurs a point mutation within the β-amyloid region of APP causing a Glu to Gly alteration; cerebrovascular amyloidosis and diffuse plaques are characteristic of this disorder.[22,23] In individuals with Down syndrome (DS) the neuropathological changes of AD occur in the brain by age 40. The disorder has been attributed to the additional genes resulting from the triplication of chromosome 21, including the APP gene; in these cases diffuse β-amyloid-containing plaques occur in neocortex and are present in the brain as early as age 10.[24,25]

APOLIPOPROTEIN E

In biochemical investigations it was observed that a small number of proteins in CSF were capable of binding to APP, one of which was identified as apolipoprotein E (ApoE), a principal transporter of cholesterol.[26] A direct relationship between ApoE and β-amyloid revealed an allele-specific interaction. ApoE4 bound soluble beta amyloid peptide more avidly than did ApoE3; further, antibodies to ApoE stain amyloid deposits in AD and other amyloidoses.[27] These data provided a connection between β-amyloid and ApoE with respect to this aspect of the molecular pathogenesis of AD.

Apolipoprotein E (ApoE) occurs on chromosome 19.[27] It appeared significant that this locus was near the genetic markers showing linkage and/or association with late-onset familial AD (FAD).[27] Subsequent genetic analyses on sporadic AD cases and early and late onset FAD probands revealed that the E4 allele of the ApoE gene was strongly associated with both sporadic AD and with late onset FAD but not with early onset FAD.[28] ApoE alleles were also studied in autopsy confirmed AD patients from 30 families with multiple affected members (28 had late onset after age 60) and nondemented controls. There was a higher frequency of ApoE4 in AD (52%) than in nondemented controls (16%).[26] It was found that the ApoE4 allele is present in over 65% of the cases with late onset FAD and 50% of the cases of sporadic AD with onset between ages 65-80, whereas it is present in only 30% of controls.[29] It was further observed that the ApoE4/E4 allele that is normally present in 3% of the population increases the odds of getting AD 7-fold; the ApoE4/E3 alleles, which is present in 23% of the population increases the odds of getting AD 3-4 times.[29]

The mechanism by which ApoE4 causes an effect in AD is not known with certainty at this time. However, ApoE4 may promote the formation of insoluble fibrils from β-amyloid, which as noted previously have neurotoxic properties. The involvement of ApoE4 with β-amyloid in AD suggests that a unified mechanism may be responsible for the formation of senile plaques. The aforementioned studies strongly argue for therapeutic investigations that take into account both gene products.

ANTISENSE OLIGONUCLEOTIDES

Antisense oligonucleotides are synthetic pieces of single stranded DNA that hybridize to targeted mRNA via nucleotide base-pairing to inhibit expression of specific proteins. The primary advantage of ASOs, relative to many other potential therapeutic agents, is their specificity: an oligonucleotide comprised of 17 nucleotides or more could, in principle, inhibit any unique gene within the human genome. Whereas intracellular nucleases degrade oligonucleotides containing phosphodiester linkages, ASOs with a phosphorothioate backbone, in which one of the nonbridged oxygen atoms is replaced by sulfur, are significantly more stable after introduction into cells and tissues. Structural and functional studies on oligodeoxynucleotide pohosphorothioates were described in previous reports.[30-32]

The aim of the current research was to test the potential behavioral toxicity of phosphorothioate-linked ASOs directed against the human and rat APP and the human ApoE mRNA in rats. As indicated in the Introduction to this chapter, the rationale for examining behavioral toxicity was that, when considering new therapeutic approaches for the treatment AD, it is essential to avoid the administration of compounds that may further compromise cognition and memory. We addressed this issue by designing a multi-component water maze paradigm that was used to assess behavioral toxicity.

SPATIAL TASKS DESIGNED TO TEST LEARNING AND MEMORY

Both theoretical[33] and empirical evidence[34-38] suggests that spatial tasks have the capacity to test three forms of learning—specifically place, cue and response learning. In place learning, the topography of multiple environmental stimuli defines a spatial location. In utilizing place learning, an animal navigates by using the configurational relationships among the stimuli that surround a goal. An example of human place learning would be an individual who parks a car in an empty parking lot, only to return to find the lot full. Despite the fact that the car is hidden from view, the person may nevertheless walk directly to the car by using both the angular relationships and distances among the stimuli that surround the lot. In cue learning, a single stimulus evokes a general approach response, which leads to a larger or more intense version of the same stimulus. In using cue learning, an animal navigates down a stimulus gradient towards a goal characterized by sensory qualities. A familiar example of human cue learning would be the person who, from a distance, approaches an "Exit" sign in order to leave a building. Finally, in response learning, a stimulus evokes a more specific motor act, such as a turn, which usually leads to a different stimulus. In utilizing response learning, an animal navigates towards a goal on the basis of a turn or a sequence of turns. An example of human response learning would be the person who makes a left turn in their car in the presence of a "Left Turn Only" sign.

WATER MAZE PARADIGM FOR ASSESSING LEARNING AND MEMORY

Since its introduction by Morris,[34] the water maze has become an important apparatus for administering spatial tasks to the rat. The maze consists of a large circular tank, filled with cold water made opaque by the addition of nontoxic dye. Within the maze, a platform is made available for escape from the cold water and the exertion of swimming.

The water maze offers several advantages over other test apparatuses. First, no deprivation procedures are needed to induce the motivation to escape. Second, in the rat, swimming is a natural behavior that requires no shaping procedures. Third, the water maze constitutes a "forced-choice" apparatus in which rats are not given the option of not responding. Finally, the water maze possesses a high degree of flexibility. Thus, a number of experimental variables can be manipulated to isolate and examine place, cue or response learning:

Environment—Studies can be perfomed within environments that vary in the availability of extra-maze stimuli. Generally, tasks conducted in environments containing multiple extra-maze stimuli encourages place learning. In contrast, those conducted in environments lacking extra-maze stimuli encourage either cue or response learning, depending on the visibility or invisibility of the goal, respectively.[33,34,36]

Platforms—Two types of platforms can be provided for escape: (1) a submerged platform, resting under the surface of the water and painted the same color; or (2) a visible platform, projecting from the surface and painted a contrasting color. The submerged platform is used to study either place or response learning, while the visible platform is employed to study cue learning.[34,36,39]

Platform Mobility—Place, cue and response learning can be further subdivided on the basis of the temporal factors. In animals, two temporal levels of memory are generally recognized: reference memory, which stores information on the order of hours, days or years, and is roughly equivalent to human long-term memory; and working memory, which stores information on the order of seconds or minutes, and approximates human short-term memory.[40]

Manipulating the mobility of the submerged platform is an important procedure for differentiating reference and working place memory. Typically, two mobility procedures are utilized with the submerged platform. First, the platform can remain in a single location both within, and across, test sessions (Fixed-Location Test Procedure). Although this is the most popular test procedure, it fails to differentiate between reference and working place memory. For example, if a rat swims directly from a starting point to the submerged platform, the performance may reflect memory of the location trained during previous test sessions (reference place memory), or it may reflect memory of the location trained during the current test session (working place memory). Second, the platform can remain in a single location within a test session, but be re-positioned to a different location on the following session (Variable-Location Test Procedure). In this procedure, reference place memory contains information about the entire set of previously-trained locations. In contrast, working place memory contains information about the currently-trained location. Rats typically use reference place memory during the first trial of a session to explore previously-trained locations in an attempt to discover the currently-trained location, and working place memory during the latter trials of a session to return directly to the currently-trained location (Binsack and Treichler, 1995, submitted for publication).

The manner in which the environment, platform type and platform mobility are combined produces both a single-component and a multi-component approach to water maze research. In the single-component approach, experimental variables are tightly controlled in order to restrict the rat to a single form of learning. This approach requires relatively few dependent measures to assess whether the form of learning under consideration is affected by a treatment. In the multi-component approach, the experimental variables are not as tightly controlled. Consequently, the rat may display several forms of learning within the same task. While the multi-component approach saves time and effort, it requires a variety of dependent measures to distinguish which forms of learning are affected by a particular treatment. Typically, we use three dependent measures, analyzed both across and within test sessions, to assess learning:

> Turning Bias—As a standard procedure, rats are placed in the water maze facing the wall.[34] This orientation forces each rat to execute a left or right turn at the start of a trial in order to face the interior of the maze, and ultimately, the escape platform. However, although rats can turn in either direction, over the course of training they invariably develop a preferred direction.[41] Because Turning Bias examines the acquisition of a specific motor act, it reflects response learning.
>
> To determine Turning Bias, the direction each rat turns at the start of a trial is recorded. At the end of training, the preferred direction is determined for each rat by simple summation. Subsequently, for each rat, each individual trial is scored as starting with a turn in either the preferred direction (+1) or the nonpreferred direction (0). Finally, motor learning is expressed as the probability of executing the preferred turn.
>
> Heading Errors—After the turn, the rat has two behavioral options, depending on the type of platform used on a trial: if it has learned the location of the submerged platform, or has learned to approach the visible platform, the rat takes a direct route towards it; conversely, if it fails to learn the location of the submerged platform, or fails to learn to approach the visible platform, the rat swims a more circuitous route. Because Heading Errors examine the memory for a location or a visual stimulus, they reflect place and cue learning, respectively.

Heading Errors are determined for each trial by constructing a hypothetical alley, one body-length wide, from the start point to the platform.[36] If the rat remains inside the alley, it is judged to have navigated directly to the platform, and no error is recorded for that trial (0); conversely, if the rat swims outside the alley, it is judged to have taken an indirect route, and an error is recorded (+1). Place and cue learning is exhibited as a decreased probability for committing Heading Errors during submerged and visible platform trials, respectively.

Latency—Latency reflects the time taken to reach the platform after release. While undoubtedly the most popular dependent variable used in water maze research, latency is only an indirect measure of learning. Specifically, when latency is the only dependent measure used to gauge performance, it is impossible to differentiate between a treatment that disrupts learning from one that slows swim speed.

Because other accuracy measures such as Turning Bias and Heading Errors are better able to assess place, cue and response learning, we typically use latency measures to gauge the effect of a treatment on motor speed.

SENSITIVITY OF WATER MAZE TO COMPOUNDS WITH ANTICHOLINERGIC ACTIVITY

In considering a therapeutic intervention for AD we have been concerned with the need to evaluate new compounds for their potential anticholinergic effects. To our knowledge, ASOs were not previously evaluated for their potential capacity to interfere with acetylcholine metabolism for functioning. The reader is referred to ref. 42 for a discussion of drug categories under consideration for AD with emphasis on the cholinergic system. Prior to testing ASOs in the water maze paradigm, the sensitivity of the test system for detecting anticholinergic activity, as expressed in a memory deficit, was assessed.

Figure 13.1 represents the results from a completed multi-component study designed to simulate cholinergic denervation in AD brains;[43] detailed data will be presented elsewhere (Binsack and Treichler, 1995, submitted for publication.) The data of Figure 13.1 illustrates the efficiency of utilizing a multi-component task. The primary aim of the study was to examine the effect of cholinergic-muscarinic blockade on working place memory; a secondary aim was to assess—simultaneously—the effect of muscarinic blockade on cue and/or response learning. Rats were injected with either atropine sulphate ($AtSO_4$), a central and peripheral muscarinic receptor blocker; atropine methylnitrate ($AtMeNO_3$), a peripheral muscarinic blocker; or saline. Training consisted of 12 sessions, with each session composed of 8 trials. Within each session,

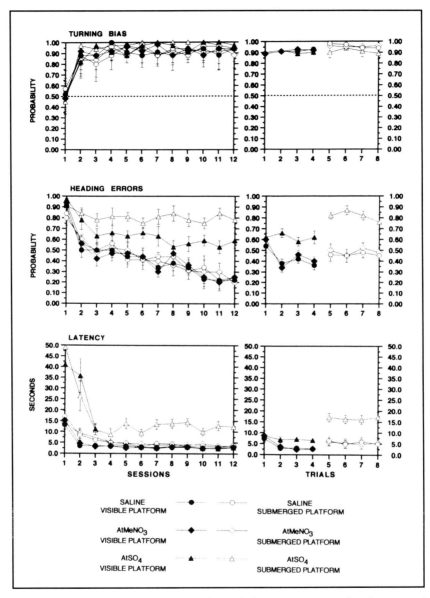

Fig. 13.1. Water maze assessment of anticholinergic compounds. The general conditions for testing, as described in the text. Abbreviations used: $AtSO_4$, atropine sulphate, $AtMeNO_3$, atropine methylnitrate.

the rats were given two types of trials: during the first 4 trials, the visible platform was used for escape; during the last 4 trials, the submerged platform was used, and was positioned in the same location. Across test sessions, the platforms were re-positioned to a different location. Throughout visible and submerged platform trials, the rats had free access to extra-maze stimuli.

Figure 13.1 (top) illustrates Turning Bias measures for all three groups. Across sessions (Fig. 13.1; top left), there was a significant increase in turning probabilities in each drug group, revealing that all rats learned to execute a turn in a preferred direction at the start of a trial. There were no differences between drug groups, showing that muscarinic blockade had no effect on the rate of acquisition or the asymptotic performance of the directional turn. Within each group, there were no differences between visible and submerged platform trials, suggesting that all rats executed the directional turn on both types of trials. Within sessions (Fig. 13.1; top right), there were no changes in directional turning probabilities, suggesting that response learning was mediated by reference memory.

Figure 13.1 (middle) illustrates Heading Error results. Across sessions (Fig. 13.1; middle left), AtMeNO$_3$ and Saline rats exhibited a gradual decrease in Heading Errors during both visible and submerged platform trials, implying that peripheral muscarinic blockade had no effect on either cue learning or place learning, respectively. Within sessions (Fig. 13.1; middle right), the AtMeNO$_3$ and Saline rats showed a significant decline in Heading Errors between the first and second visible platform trials, with no differences among either the remainder of the visible trials (Trials 2-4) or the submerged trials (Trials 5-8). The latter results suggest not only that the AtMeNO$_3$ and Saline groups learned to navigate to the visible platform (cue learning), but that they simultaneously learned the location of the visible platform (working place memory) and used the memory to navigate to the submerged platform on subsequent trials. In contrast, the AtSO$_4$ rats displayed no improvement in Heading Errors during submerged platform trials, implying that central muscarinic blockade significantly impaired working place memory.

The effect of AtSO$_4$ on cue learning was more difficult to determine. Because the visible platform trials were conducted in the presence of extramaze cues, the visual properties of the platform encouraged cue learning, while the configuration of extramaze cues encouraged place learning. Nevertheless, two lines of evidence suggest that AtSO$_4$ had no effect on cue learning per se, but instead, disrupted place learning intrinsic to the visible platform trials. First, if AtSO$_4$ had completely disrupted cue learning, then the visual properties of the visible platform should have provided no advantage over the submerged platform. Thus, Heading Error probabilities during visible and submerged platform trials would have been equivalent. However, this is clearly not the case: AtSO$_4$ rats committed fewer Heading Errors during visible platform trials than during submerged platform trials. Second, if AtSO$_4$ had only partially disrupted cue learning, then rats should commit more Heading Errors than controls during visible platform trials—irrespective of the platform's mobility. However, although the AtSO$_4$ rats in the current study clearly committed a greater number of Heading Errors than controls when tested under variable-location procedures, AtSO$_4$ rats in previous studies have failed to commit a greater number of Heading Errors when tested under fixed-location procedures.[36] The results suggest that re-positioning the visible platform in each session disrupted the ability of AtSO$_4$ rats to locate (place learning), but not necessarily to approach (cue learning), the platform.

Finally, Figure 13.1 (bottom) represents Latency results. As might be expected on the basis of Heading Errors, AtSO$_4$ rats took longer than AtMeNO$_3$ and Saline rats to escape during the submerged platform trials. Unexpectedly, AtSO$_4$ rats failed to show longer Latencies during visible platform trials, confirming the observation that AtSO$_4$ rats swam faster than Controls.

Together, the outcomes reveal that central, but not peripheral, muscarinic blockade impaired working place learning while sparing cue and response learning. Moreover, because the AtSO$_4$ rats exhibited cue and response learning indistinguishable from controls, place learning impairments cannot be attributed to nonspecific performance effects such as alterations in visual acuity, motivation, attention, or swim speeds. These data support the use of the water maze paradigm for assessing potential anticholinergic activity of pharmacological compounds.

CONDITIONS FOR TESTING ANTI-APP AND ANTI-APOE ASOS

Subjects were male Long-Evans rats, 60-90 days old at the start of testing, with weights ranging between 300-525 gm. Rats were housed in clear plastic boxes, and had access to Purina lab chow and water ad libitum.

A circular galvanized steel tank served as the water maze. The walls and floor of the maze were painted flat white. Titanium dioxide (Dupont), a nontoxic powder, was added to each tank of water to render it opaque white. A video camera was suspended above the center of the maze and fed into a TV monitor and VCR for remote viewing and taping. Throughout testing, subjects had free access to numerous extra-maze stimuli, consisting of electronic equipment, shelving, doors, sinks, etc.

Oligodeoxynucleotide phorphorothioates were synthesized by using the β-cyanoethyphosphoramidite approach.[44,45] The following sequences were synthesized: a 24 mer phosphorothioate deoxyribonucleotide complementary to the initiation codon region of rat APP mRNA (ANTI-(R)-APP);[46] a 24 mer with a sequence complementary to the initiation codon region of human APP mRNA (ANTI-(H)-APP (8); a 27 mer complementary to the initiation codon of human apolipoprotein-E (ANTI-(H) APO-E) mRNA;[48] and a 24 mer control compound with a randomized oligonucleotide sequence (NONSENSE). Each oligonucleotide was dissolved in physiological saline at room temperature, proportioned into aliquots and kept frozen until used. Rats were assigned to 1 of 5 treatment groups: Anti-(H)-APP, Anti-(R) APP, Anti-(H)-APO-E, Nonsense and Saline in which rats were given equivalent volumes of saline. Within each treatment group, the rats were administered either 2.5 mg/kg or 5.0 mg/kg of oligonucleotide. To administer the compounds, the rats were placed in a jar and lightly anesthetized with Metofane. After anesthetization, the tail vein was located, and the rat was injected via a 26-gauge needle over the course of 5 min.

Previous pharmacokinetic research indicated that while ASOs have a limited absorption in brain, they have a relatively low rate of elimination.[47] After an initial 70% reduction in the maximal ASO concentration over the first 2 hours, approximately 30% is detectable from 3 to 240 hours after injection. Consequently, rats did not require daily injections to maintain the steady state level of oligonucleotide. Over several days, each rat was subjected to 3 injection-test cycles, consisting of one injection session followed 24 hours later by four behavioral test sessions. Thus, injections were given on days 1, 6, and 11, with behavioral tests given on days 2 through 5, 7 through 10, and 12 through 15.

The water maze was divided into four quadrants, NW, NE, SW, SE, with the center of each designated as a platform location. Training consisted of 12 test sessions, with each session composed of 12 trials. Within each test session, the rats were given two types of trials: during the first 8 trials, the submerged platform was used for escape; during the last 4 trials, the visible platform was used, and was positioned in the same location. Each trial started from one of four start points located around the perimeter of the water maze (N, S, E, or W), with each start point used three times in a session. Individual trials began by placing the rat in the water, facing the wall, at a start point. If the rat found the platform, it was allowed to remain there for 30 sec; if it failed to find the platform within 60 sec, it was guided to the platform by hand. Across test sessions, the platforms were repositioned to different locations. Unlike the $AtSO_4$ study, the submerged platform trials preceded the visible platform trials. This manipulation afforded the opportunity to examine both reference and working place memory in addition to cue and response learning.

While testing of the described type is still in the early stages of development and application, an estimate of the effects of oligonucleotides on learning and memory can be obtained by examining the means of the dependent measures taken throughout testing. Figures 13.2, 13.3 and 13.4 represent the water maze performance of the ANTI-APP, ANTI-APO-E, and NONSENSE groups of rats.

Figure 13.2 (top) represents Turning Bias results for rats treated with ANTI-APP or Saline. Across test sessions (Fig. 13.2; top left), all groups exhibited an increase in directional turning probabilities, re-

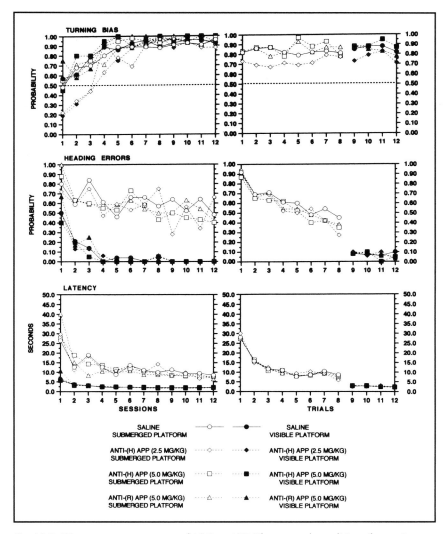

Fig. 13.2. Water maze assessment of ASO to APP. The general conditions for testing are described in the text. Abbreviations used: ANTI-(H)-APP, antisense to human APP; ANTI (R)-APP, anti-sense to rat APP.

vealing that all rats learned to execute a specific turn at the start of each trial. Thus, all rats used response learning in order to orient towards the interior of the water maze. There was considerable overlap in turning probabilities during submerged and visible platform trials, indicating that the type of platform used on a particular trial had no influence on the direction of the turn.

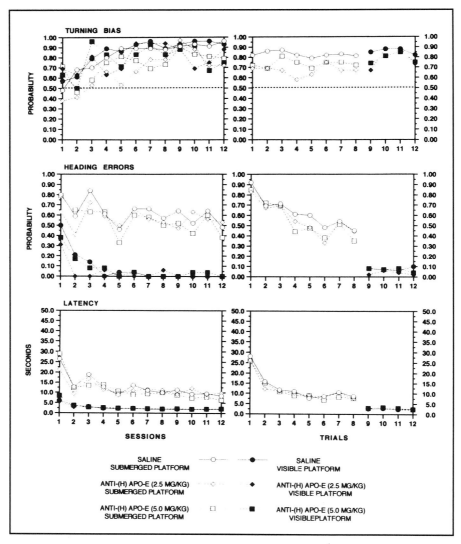

Fig. 13.3. Water maze assessment of ASO to apolipoprotein E. The general conditions for testing, as described in the text. Abbreviation used: ANTI-(H)-APO-E: antisense to human apolipoprotein E.

Within sessions (Fig. 13.2; top, right), none of the treatment groups exhibited a change in directional turning probabilities during either submerged or visible platform trials. As in the AtSO$_4$ study, the flat turning probabilities suggested that response learning had been mediated by reference memory.

Figure 13.2 (middle) illustrates Heading Errors for the Anti-APP and Saline groups. All three groups committed fewer Heading Errors during visible platform trials than during submerged platform trials. The differential performance reveals that the rats had learned to approach the visible platform—an indication of cue learning. Across sessions (Fig. 13.2; middle, left), there seem to be no differences between Anti-APP and Saline groups, suggesting that the ASO had no effect on the early acquisition (Session 1 to Session 3) or asymptotic performance (Session 4 to Session 12) of cue learning. Simultaneously, Anti-APP and Saline groups showed a gradual reduction in Heading Errors during submerged platform trials, suggesting that all rats had learned to use the configuration of extra maze cues to swim directly towards the submerged platform—an indication of place learning. There were no consistent Heading Error differences between Anti-APP and Saline groups, indicating that the ASOs had no effect on place learning. Finally, the session-to-session reduction rates for submerged platform trials seem more gradual than that for the visible platform trials, suggesting that place information was more difficult to learn than cue information.

Within sessions (Fig. 13.2; middle, right), Anti-APP and Saline groups exhibit a sharp reduction in Heading Errors during submerged platform trials, revealing an increased specificity for the platform location. The groups do not seem to differ on the first one or two trials, when rats typically use reference place memory to search previously-trained platform locations for the currently-trained location. Thus, ASOs do not seem to disrupt reference place memory. Further, the groups do not seem to differ on the remainder of the submerged platform trials, when rats utilize working place memory to return directly to the currently-trained location. Thus, the phosphorothioate oligonucleotides did not appear to disrupt working place memory. None of the groups showed any change in Heading Errors during visible platform trials, suggesting that, like response learning, cue learning had been mediated by reference memory. Figure 13.2 (bottom) represents Latencies for the Anti-APP and Saline groups. In general, the results paralleled Heading Error scores. Both across (Fig. 13.2; bottom, left) and within (Fig. 13.2, bottom, right) sessions, there seemed to be no differences among groups, indicating that ASOs, either human or rat, had no effect on swim speed.

Figure 13.3 reflects results after administration of Anti-(H)-APO-E or Saline to rats, while Figure 13.4 shows data obtained from rats in the NONSENSE category. As can be seen, none of the treatment groups, at any of the dosages tested thus far, show any consistent differences compared to the SALINE group. Overall there were no differences between ANTI-APP (human or rat sequences) or ANTI-APO-E, and NONSENSE or SALINE groups, indicating that the compounds did not disrupt reference cue memory at dosages thus far examined.

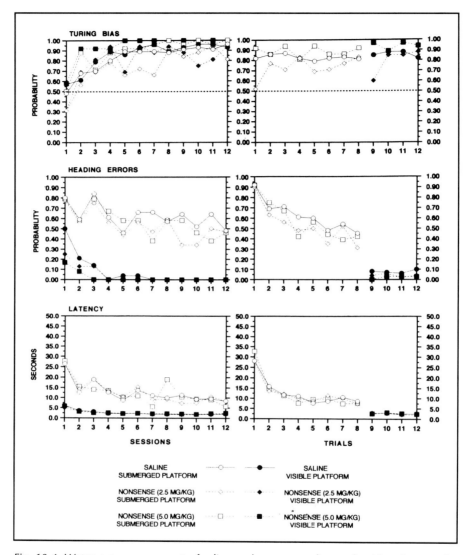

Fig. 13.4. Water maze assessment of saline and nonsense oligonucleotides. See text for experimental conditions and interpretations. The general conditions for testing, as described in the text.

FUTURE STUDIES

We have used a novel approach to the characterization of antisense oligonucleotides designed as potential therapeutic agents for the treatment of AD. Potential drug side effects, in particular as they may affect cholinergic transmission, require a fine level of assessment in appropriately designed animal paradigms. The water maze protocols described here are expected to find widespread usefulness for assessing the behavioral consequences of potential drugs when the target disease involves deterioration of cognition and memory. Concurrent with these investigations we have begun to explore genetically modified rats that overexpress an AD-related gene, the human APP_{770} (Binsack, DP, Marotta CA, in preparation). In this instance the water maze is used to assess the baseline performance of the mutants and then the effect of specific ASOs designed to halt the corresponding mRNA. As we obtain increased insights into the genetic basis of neurodegenerative and neuropsychiatric disorders, genetically modified animals will become more readily available to elucidate the nature of the disease and for evaluating new drugs. The application of water maze protocols designed to measure selective aspects of behavior combined with highly specific ASOs promises to be a fruitful approach to creating new therapeutic initiatives for a wide variety of disorders.

REFERENCES

1. McKhann G, Drachman D, Folstein M, Katzman R. Price D, Stadian EM. Clinical diagnosis of Alzheimer's disease: report of the NINCDS-ADRDA Work Group under the auspices of Department of Health and Human Serviices Task Force on Alzheimer's Disease. Neurology 1984; 34:939-944.
2. Cummings JL, Benson DF. Dementia: a clinical approach, 2nd edition. Boston: Butterworth-Heinemann, 1992:9.
3. Khachaturian, ZS. Diagnosis of Alzheimer's disease. Arch. Neurol 1985; 42:1097-1105.
4. Tierney MC, Fisher RH, Lewis AJ, Zorzitto ML, Snow WG, Reid DW, Nieuwstraten P. The NINCDS-ADRDA Work Group criteria for the clinical diagnosis of probable Alzheimer's disease: a clinicopathologic study of 57 cases. Neurology 1988; 38:359-64.
5. Mirra SS, Heyman A, McKeel D, Sumi SM, Crain BJ, Brownlee LM, Vogel FS, Hughes JP, van Belle G, Berg L. The Consortium to Establish a Registry for Alzheimer's Disease (CERAD). Part II. Standardization of the neuropathologic assessment of Alzheimer's disease. Neurology 1991; 41:479-86.
6. Katzman R, Terry R, DeTeresa R, Brown T, Davies P, Fuld P, Renbing X, Peck A. Clinical, pathological, and neurochemical changes in dementia: a subgroup with preserved mental status and numerous neocortical plaques. Ann Neurol 1988; 23:138-44.
7. Terry RD, Gonatas NK, Weiss M. Ultrastsructural studies in Alzheimer's presenile dementia. Am J Pathol 1964; 44:269-297.
8. Kang, J, Lemaire, JG, Unterbec, Salbaum JM, Masters CL, Grzeschik KH, Multhaup G, Beyreuther K., Muller-Hill B. The pecursor of Alzheimer's disease amyloid A4 protein resembles a cell-surface receptor. Nature 1987; 325:733-736.

9. Tanzi RE, McClatchey AI, Lamperti ED, Villa-Komaroff L, Gusella JF, Neve RL. Protease inhibitor domain encoded by an amyloid protein precursor mRNA associated with Alzheimer's disease. Nature 1988; 331:528-530.
10. Yankner BA, Duffy LK, Kirschner DA. Neurotrophic and nekurotoxic effects of amyloid β protein: reversal by tachykinin neuropeptides. Science 1990; 250:279-282.
11. Pike, CJ, Walencewicz, AJ, Glabe CG, Cotman CW. In vitro aging of β-amyloid protein causes peptide aggregation and neurotoxicity. Brain Res 1991; 563:311-314.
12. Busciglio J, Lorenzo A, Yankner BA. Methodological variables in the assessment of beta amyloid neuorotoxicity. Neurobiol Aging 1992; 13:609-612.
13. Emre M, Geula C, Ransil BJ, Mesulam M-M. The acute neurotoxicity and effects upon cholinergic axons of intracerebrally injected β amyloid in rat brain. Neurobiol Aging 1992; 13:553-560.
14. Kowall, NW, McKee, AC, Yankner, BA, Beal, MF. In vivo neurotoxicity of β-amyloid in rat brain. Neurobiol Aging 1992; 13:553-560.
15. Hendriks L ,van Duijn CM, Cras P, Cruts M, Van Hul W, van Harskamp F, Warren A, McInnis MG, Antonarakis SE, Martin JJ et al. Presenile dementia and cerebral haemorrhage linked to a mutation at codon 692 of the beta-amyloid precursor protein gene. Nat Genet 1992; 1:218-21.
16. Mullan M, Crawford F, Axelman K, Houlden H, Lilius L, Winblad B, Lannfelt L. A pathogenic mutation for probable Alzheimer's disease in the APP gene at the N-terminus of beta-amyloid. Nat Genet 1992; 1:345-347.
17. Chartier-Harlin MC, Crawford F, Houlden H, Warren A, Hughes D, Fidani L, Goate A, Rossor M, Roques P, Hardy J et al. Early-onset Alzheimer's disease caused by mutations at codon 717 of the beta-amyloid precursor protein gene. Nature 1991; 353:844-846.
18. Goate A, Chartier-Harlin MC, Mullan M, Brown J, Crawford F, Fidani L, Giuffra L, Haynes A , Irving N, James L et al. Segregation of a missense mutation in the amyloid precursor protein gene with familial Alzheimer's disease. Nature 1991; 349:704-706.
19. Murrell J, Farlow M, Ghetti B., Benson MD. A mutation in the amyloid precursor protein associated with hereditary Alzheimer's disease. Science 1991; 254:97-99.
20. Cai X-D, Gold TE, Younkink SG. Release of excess amyloid β protein from a mutant amyloid β protein precursor. Science 1993; 259:514-516.
21. Citron M, Oltersdorf T, Haass C, McConlogue L, Hung AY, Seubert P, Vigo-Pelfrey C, Lieberburg I, Selkoe DJ. Mutation of the beta-amyloid precursor protein in familial Alzheimer's disease increases beta-protein production. Nature 1992; 360(6405):672-4.
22. Levy E, Carman MD, Fernandez-Madrid IJ, Power MD, Lieberburg I, van Duinen SG, Bots GT, Luyendijk W, Frangione B. Mutation of the Alzheimer's disease amyloid gene in hereditary cerebral hemorrhage, Dutch type. Science 1990; 248:1124-1126.

23. Van Broeckhoven C, Haan J, Bakker E, Hardy JA, Van Hul W, Wehnert A, Vegter-Van der Vlis M, Roos RA. Amyloid beta protein precursor gene and hereditary cerebral hemorrhage with amyloidosis (Dutch). Science 1990; 248:1120-1122.
24. Wisniewski KE, Wisniewski H.M, Wen GY. Occurrence of neuropathological changes and dementia of Alzheimer's disease in Down's syndrome. Ann Neurol 1985; 17:278-282.
25. Rumble B, Retallack R, Hilbich C, Simms G, Multhaup G, Martins R, Hockey A, Montgomery P, Beyreuther K, Masters CL. Amyloid A4 protein and its precursor in Down's syndrome and Alzheimer's disease. N Engl J Med 1989; 320:1446-452.
26. Strittmater WJ, Saunders AM, Schmechel D et al. Apolipoprotein E: high avidity binding to β-amyloid and increased frequency of type 4 allele in late-onset familial Alzheimer disease. Proc Natl Acad Sci USA 1993; 90:1977-1981.
27. Pericak-Vance MA, Bedout JL, Gaskell PC, Roses AD. Linkage studies in familial Alzheimer disease-evidence for chromosome 19 linkage. Am J Hum Genet 1991; 48:1034-1050.
28. Saunders AM, Strittmatter WJ, Schmechel D, George-Hyslop PH, Pericak-Vance MA, Joo SH, Rosi BL, Gusella JF, Crapper-MacLachlan DR, Alberts MJ et al. Association of apolipoprotein E allele epsilon 4 with late-onse familial and sporadic Alzheimer's disease. Neurology 1993; 3:1467-1472.
29. Roses AD, Saunders AM, Strittmatter WJ, Pericak-Vance MA, Schechel D. Association of apolipoprotein E allele E4 with late onset familial and sporadic Alzheimer's disease. Neurology 1993; 43(suppl. 2):A192.
30. Agrawal S, Mayrand SH, Zamecnik, Pederson T. Site-specific excision from RNA by Rnase H and mixed-phosphate-backbone oligodeoxynucleotides. Proc Natl Acad Sci USA 1990; 87:1401-1405.
31. Agrawal S, Temsamani J, Tang JY. Pharmacokinetics, biodistribution, and stability of oligodeoxynucleotide phosphorothioates in mice. Proc Natl Acad Sci, USA 1991; 88:7595-7599.
32. Agrawal S, Sarin PS, Zamecnik M, Zamecnik PC. Cellular uptake and anti-HIV acitivity of oligonucleotides and their analogs. In: Erickson RP, Izant JG, eds. Gene Regulation: Biology of Antisense RNA and DNA. New York: Raven Press, 1992:273-283.
33. O'Keffe J., Nadel L. The Hippocampus As A Cognitive Map. Oxford: Oxford U. Press, 1976.
34. Morris, RGM. Spatial localization does not require the presence of local cues. Learning Motivation 1981; 12:239-260.
35. Okaichi H. Performance and dominant stagegies on place and cue tasks following hippocampal lesions in rats. Psychobiol 1987; 15:58-63.
36. Whishaw IQ. Cholinergic blockade in the rat impairs locale but not taxon strategies for place navigation in a swimming pool. Behav Neurosci 1985; 5:979-1005.
37. Buresova O, Bolkhuis JJ, Bures J. Differential effects of cholinergic blockade on the performance of rats in the water tank navigation task and in a radial water maze. Behav Neurosci 1986; 100:476-482.

38. Ellen P, Taylor HS, Wages C. Cholinergic blockade effects on spatial integration versus cue discrimination performance. Behav Neurosci 1986; 100:720-8.
39. Sutherland RJ, Chow GL, Baker JC, Linngard RC. Some limitations on the use of distal cues in place navigation by rats. Psychobiol 1987; 15:48-57.
40. Olton DS, Samuelson, RJ. Remembrance of places past: spatial memory in rats. J Exp Psychol 1979; 2:97-116.
41. Whishaw IQ, Tomie J. Cholinergic blockade produces impairments in a sensorimotor subsystem for place navigation in the rat: evidence from sensory, motor and acquisition tests in a swimming pool. Behavioral Neurosci 1987; 101:603-616.
42. Growdon JH. Biologic therapies for Alzheimer's Disease. In: White, PJ, ed. Dementia. Philadelphia: FA Davis, 1993:375-399.
43. Geula C, Mesulam M-M. Cholinergic systems and related neuropathological predilection patterns in Alzheimer Disease. In: Terry RD, Katzman R, Bick KL eds. Alzheimer Disease. New York: Raven Press, 1994:263-293.
44. Beaucage, SL. In: Agrawal, ed. Protocols for Oligonucleotides and Analogs (Methods in Molecular Biology Series). New Jersey: Humana Press, 1993:33-61.
45. Padmapriya AA, Tang J, Agrawal S. Large-scale synthesis, purification, and analysis of oligodeoxynucleotide phosphorothioates. Antisense Res Dev 1994; 4:185-99.
46. Chernak JM. Structural features of the 5' upstream regulatory region of the gene encoding rat amyloid precursor protein. Gene 1993; 133:255-260.
47. Zhang R, Robert RB, Lu Z, Liu T, Jiang, Z, Galbraith W, Agrawal S. Pharmacokinetcs and tissue distribution in rats of an oligodeoxynucleotide phosphorothiate (GEM91) developed as a therapeutic agent for human immunodeficiency virus type-1. Biochem Pharm 1995; 49:929-939.
48. Paik YK, Chang DJ, Reardon CA, Davies GE, Mahley RW, Taylor JM. Nucleotide sequence and structure of the human apolipoprotein E gene. Proc Natl Acad Sci USA 1985; 82:3445-9.

INDEX

A

Adenovirus, 140
Adenylate cyclase, 83
Adenylyl cyclase, 54
Adipocytes, 182
α_2-Adrenoceptor agonists, 63
Adrenocorticotrophic hormone (ACTH), 154
Al Ghoul WM, 194
Alz-50, 193
Alzheimer's disease (AD), 181, 194, 201-202, 216-217
   antisense oligodeoxynucleotides, 204-205
      water maze
         amyloid precursor protein (APP), 211-215, 213f, 216f
         apolipoprotein E (ApoE), 211-215, 214f, 216f
Amphetamine, 83, 115, 116
Amygdala, 167
β-Amyloid, 202-203
Amyloid precursor protein (APP), 213f. See also β-Amyloid.
Antinociception, 26, 27, 39-42, 40f, 41f, 43f, 44, See also Opioid receptors.
   nonopioid, 63
Antisense oligodeoxynucleotides (ODN), 112-114
   delivery, 139-140, 164, 164f, 176
      plasma membrane permeabilization, 141, 142f
      vector/conjugate systems, 140-141
      whole cell patch clamp technique/microinjection techniques, 143
   inhibition of receptor expression, 2t
      methodology, 3-6, 3t
   knockout, in vivo, 96
   mechanisms of action, 136-139, 136f, 138f, 139f
Anxiety, 153-154
   corticotropin-releasing hormone (CRH) and antisense
   oligodeoxynucleotides, 158, 159f, 160, 161f, 162-163
   receptor type I, 163-167, 164f
AP1 (activator protein 1), 112, 117, 120, 122
Apolipoprotein E (ApoE), 203-204
Apomorphine, 99, 100f, 101f, 104, 115, 116
Apoptosis, 122
Arginine vasopressin (AVP), 156

B

Baroreceptor sympathetic reflex, 118
Basal ganglia
   Fos in vivo, 114-118
bcr/abl, 183
Benzodiazepines, 154
Beta-adrenergic receptor kinase, 179
B-HT920, 66t
Bilsky EJ, 42
Blood pressure control
   Fos, 118-119
Britton DR, 162
Brain
   ODN introduction into, 5
Butler PD, 162
BW37373U86, 39

C

Ca^{2+}
   channels
      voltage-activated (VACC), 144-145
   neuronal death, 195
Caceres A, 192
Caine SB, 84
Calcitonin, 163
cAMP-dependent protein kinase. See Protein kinase A (PKA).
Carbachol, 138f
Casein kinase 2 (CK2), 176-178, 177f, 178f, 179f
Central nucleus (CeN), 167
c-fos, 73, 82-83
c-fos, 112
Chalmers JP, 118
Chiasson BJ, 115, 117, 118, 120
Chien C-C, 41
Chiu TH, 119
Chlordiazepoxide, 162
Cholera toxin (CTX), 58
Cholesterol, 140
Circadian rhythm and sleep/waking
   Fos, 119-120
Cirelli C, 120
Clonidine, 66t
Cocaine, 72, 83, 84, 115, 116
Collins KA, 42
Corticosterone, 156, 157
Corticotropin-releasing hormone (CRH), 154
   antisense oligodeoxynucleotides, 155-156, 156t, 157f, 158
   anxiety, 158, 159f, 160, 161f, 162-163
Corticotropin-releasing hormone (CRH) receptor type I
   anxiety, 163-167, 164f
CRE (cyclic-AMP-response element), 117
CREB (cyclic-AMP-response element binding protein), 116, 117, 122
Cyclin-dependent protein kinases (CDKs), 181
   cdk1 and cdk2, 182
Cyclins, 182

D

[D-Ala², Glu⁴]deltorphin, 26, 27
DALCE ([D-Ala², leu⁵, Cys⁶]enkephalin), 26, 27
DAMGO ([D-Ala²-NMePhe⁴-Gly⁵-ol]enkephalin), 31, 38, 46-47, 47t, 58, 59, 60
DAP kinase, 183
Dopamine, 71-72, 93-94, 114, 144
   Fos, 114-115
   receptors, 72-73, 80, 80f, 95-96
      antisense oligodeoxynucleotides, 73-75, 81-85
         autoradiography and histology, 98, 101-102
         electrophysiological measurements, 97-98
         intranigral administration, 96-97
      autoreceptors, 84, 94-95, 102, 104
         D_1, 75-76, 76f, 77f, 78-80, 78f, 79f, 80f, 82-84, 83f
         D_2, 85, 99-102, 99t, 100f, 101f, 102t, 102, 103f, 104-106
         D_3, 81, 82f, 84, 102
DOR-1 clone, 13-15, 14f
Douglas WW, 135
Down syndrome, 203
DPDPE ([D-Pen², D-Pen⁵]enkephalin), 26, 27, 40, 44
Dragunow M, 82
Drug abuse, 85
Dynorphin, 45, 46f, 120

E

egr-1, 116
β-Endorphin, 42, 154
Enkephalins, 11, 120
Epidermal growth factor (EGF), 181
Exonucleases, 140
Extracellular signal regulated protein kinases (ERK). See Mitogen-activated protein (MAP) kinases.

F

Fibroblasts, 182
β-FNA (β-funaltrexamine), 26, 27, 44
Folate, 140
Fos, 112. See also c-fos.
   basal ganglia function, 114-118
   blood pressure control, 118-119
   circadian rhythm and sleep/waking, 119-120
   neuronal injury and repair, 121-122
   nociception, 120-121

G

GABA (g-aminobutyric acid), 118
GBR 12909, 72
Gillardon F, 121
Ginzburg I, 192
Glutamate, 192, 193, 194, 195, 197
Glycogen synthase kinase 3 (GSK3), 181

G-protein coupled receptors, 62
G-proteins, 53-55, 54f, 55f
   antinociception, 58
   antisense oligos, 55-56, 56f, 57f
GRE (glucorticoid response element), 117
Growth cone, 190
GTP-binding proteins, 144, 145-146
Guanfacine, 66t

H

Haloperidol, 105, 116, 117
Hanemaaijer R, 192
Heilig M, 82, 115
Heinrichs SC, 167
Henry DJ, 72
Hexosaminidase, 146
Hooper ML, 116
Hunter JC, 121
6-Hydroxydopamine, 72, 78-80, 114
Hylden JLK, 57
Hypothalamic-adrenal-pituitary system (HPA), 154, 156

I

ICI 174,864, 26, 27
Immediate early genes (IEG), 112
Initiation codon, 113

J

Jun, 112, 116, 117, 119, 120, 121, 122

K

Kalin NH, 162
Kleuss C, 144
Koob GF, 84
KOR-1, 15-16, 15f
KOR-3 antisense mapping, 17-19, 18f, 19f
Kosik KS, 192
Krox 24, 115
Kurose H, 63

L

Lactotroph anterior pituitary cells, 144
LAR, 184
Learning
   assessment in rats, 205-208
Liebsch G, 167
Liposomes, 140
Lysosomes, 140

Index

M

MAP1B (microtubule-associated protein), 176. See also Tau.
Mattson MP, 194
Medial preoptic area (MPA), 120
Megakaryocyte-associated tyrosine kinases (MATK), 183
Melanotrophs, 146
Merchant KM, 117
Methamphetamine, 72
Methylphorphonate, 5
Microtubules, 190
Miller MW, 194
Mitogen-activated protein (MAP) kinases, 182
Missense oligodeoxynucleotides, 47-48
MOR-1 antisense mapping, 16-17, 17f
Morishita R, 182
Morphine, 11, 42, 43f, 44, 58, 59, 60, 61f, 72
Morphine-6β-glucuronide (M6G), 16, 17f
Morris RGM, 205
mRNA, 13, 73, 74
Muscimol, 118, 119

N

Naloxone, 44
Naltrexone, 44
Naltriben (NTB), 27
Neural plasticity, 189-191
Neuritogenesis, 176-178, 177f, 178f, 181
Neuroblastoma cells. See NG 108-15.
Neurofibrillary triangles (NFTs), 194, 202
Neuropeptide Y-Y1 receptor, 73
Neuromedin N, 117
Neuronal injury and repair
    Fos, 121-122
Neurotensin, 116-117
NG 108-15, 29, 32
NGF (nerve growth factor), 181
NGFIA, 115, 116, 121
Nigrostratal pathways, 114
N-methyl-D-aspartate receptor, 73
Nociception
    Fos, 120-121
5'-NTII (5'-naltrindoleisothiocyanate), 26
Nucleases, 140
Nucleus accumbens, 72, 84, 115

O

7-OH DPAT (7-hydroxy-N,N-di-n-propyl-2-aminotetralin), 84
OKT3, 183
Oligodeoxynucleotides
    nonspecific effects, 74

Opioid receptors, 11-13, 12t, 19-20
    antinociception, 26, 27, 39-42, 40f, 41f, 43f, 44
    delta (δ), 11, 12, 12t, 26-28
        heterogeneity, 28-34
        DOR-1, 13-16, 14f, 15f
    kappa (κ), 11, 12, 12t
        antisense mapping
            KOR-3, 17-19, 18f, 19f
        antisense ODN design, 38-39
        KOR-1, 15-16, 15f
    knockouts, 12-13
    mu(μ), 11, 12, 12t
        antisense mapping
            MOR-1, 16-17, 17f
        antisense ODN design, 38-39
        G-protein antisense, 58-60, 59f, 60f, 62, 64f-65f
        side-effects, 62-63, 66f
    thermoregulation, 44-46, 45f, 46f
Owens MJ, 165
Oxotremorine, 79
Oxytocin, 156

P

p59fyn, 183
Paired helical filaments (PHF), 194
Parathyroid hormone, 163
Paraventricular nucleus (PVN), 154
Parkinson's disease, 114
Peptide nucleic acids (PNA), 4-5
Pertussis toxin (PTX), 58
Phosphatases, 183-184
Phospholipase D, 180, 181
Phosphorothioate oligodeoxynucleotide analogs, 1, 4-5, 5t, 113, 137, 155
Pimozide, 72
Platelet-derived growth factor (PDGF), 181
PL017 ([NMePhe³,D-Pro⁴]morphiceptin, 39, 45, 45f
polo, 183
Preoptic anterior hypothalamus, 44
Preprodynorphin, 120, 121
Proenkephalin, 117
Progesterone receptor, 73
Programmed cell death. See Apoptosis.
Proline-directed protein kinases (PDPK), 181
    cyclin-dependent protein kinases (CDKs), 181-182
    cyclins, 182
    glycogen synthase kinase 3 (GSK3), 181
    mitogen-activated protein (MAP) kinases, 182
    serine-threonine protein kinases, 182-183
Proopiomelanocorticotropine (POMC), 154
Protein kinase A (PKA), 178, 179, 180
Protein kinase C (PKC), 179, 180
    isoforms, 180-181
Protein kinase 2. See Casein kinase 2.
Protein phosphorylation, 175

Q
Quinpirole, 42, 72, 75, 77f, 79, 81, 82f, 84

R
Rab3a, 145, 146
Rab3b, 145, 146
raf kinase, 182
Rassnik S, 167
Respiratory syncitial virus (RSV), 178
RNase H, 28, 74, 113, 137
Robertson GS, 117
Rossi G, 40
Rubin RP, 135

S
SCH 23390, 72, 73
Segal D, 84
Senile plaques, 202
Serine-threonine protein kinases, 182-183
Shuttle-box conflict task, 158
SKF 38393, 72, 75-76, 76f, 77f, 78-80
Somatodendritic autoreceptors. See Dopamine, autoreceptors.
Somatostatin, 138f
Sommer W, 115
Spiperone, 105
Spiradoline, 39, 41, 43f
Standifer KM, 48
Stem cell tyrosine kinase 1 (STK-1), 183
Stimulus-secretion coupling, 135-136
Streptolysin O (SLO), 141
Stull RA, 176
Substantia nigra, 72, 84, 95
Sufentanil, 58, 59, 60
Sulpiride, 72, 81, 95
Suprachiasmic nucleus (SCN), 119
Swiergiel AH, 167

T
Tau microtubule-associated protein, 190-191, 197
    developing neurons
        antisense oligodeoxynucleotides, 192, 193t
    mature and degenerating neurons, 192-194
        antisense oligodeoxynucleotides, 194-195, 195f, 196f
Tau-1 protein kinase, See Glycogen synthase kinase 3 (GSK3).
TAU-2, 193, 193
Thyrotropin (TRH), 144
Transcription, 194
Transferrin, 140
Transgenic knockouts, 106
Tseng LF, 42, 48
Tyrosine hydroxylase, 98
Tyrosine protein kinases, 183

U
U50,488H, 41
U69,593, 31

V
Vasoactive intestinal peptide, 163
Vasopressin. See Arginine vasopressin (AVP).
Ventral striatum, 84
Ventral tegmentum, 95
Ventral tegmental-nucleus accumbens pathway, 71

W
Wahlestedt C, 47
Water maze paradigm, 205-208
    anticholinergics, 208-211, 209f
White DJ, 72
Whitesell L, 113
Wilcox GL, 57
Wollnik F, 119, 120

Z
Zamecnik PC, 1
Zhou L-W, 42